Chemistry through models

Chemistry
through models

Concepts and applications of modelling
in chemical science, technology and industry

COLIN J. SUCKLING
Royal Society Senior Research Fellow, Strathclyde University

KEITH E. SUCKLING
Lecturer, Department of Biochemistry, University of Edinburgh Medical School

CHARLES W. SUCKLING, FRS
General Manager, Research and Technology, Imperial Chemical Industries Limited

CAMBRIDGE UNIVERSITY PRESS
Cambridge
London   New York   New Rochelle
Melbourne   Sydney

Published by the Press Syndicate of the University of Cambridge
The Pitt Building, Trumpington Street, Cambridge CB2 IRP
32 East 57th Street, New York, NY 10022, USA
296 Beaconsfield Parade, Middle Park, Melbourne 3206, Australia

First published 1978
First paperback edition 1980

Printed in Great Britain at the
University Press, Cambridge

British Library Cataloguing in Publication Data

Suckling, Colin J.

Chemistry through models.

1. Chemical models
I. Title   II. Suckling, Keith E.
III. Suckling, Charles Walter
540'.22'8   QD480   77-41429
ISBN 0 521 21661 3 hard covers
ISBN 0 521 29932 2 paperback

To the supporting cast, the girls of the family:
Margaret, Karen, Catherine and Helen

CONTENTS

PREFACE

This book is a set of variations on a theme, each of which has its individual unity. The theme itself is however not merely a convenient connecting device; it is an attempt to view the practice of chemistry, both academic and industrial, in new perspective and to point out a path that may lead some way toward the solution of problems that will confront scientists in the difficult years ahead. We have, therefore, thought it best, since not everyone reads a preface, to introduce and explain our particular approach in the main text, the early paragraphs of chapter 1, rather than in this preface.

A thorough grasp of chemistry is a great asset in many jobs and we hope that this book will help to increase the recognition of modelling and of good design as necessary and powerful pieces in the chemist's tool kit, appropriate to the tackling of tasks of many kinds and we hope also to stimulate thinking on how chemists may better equip themselves to use these tools.

Six years after first considering collaborating on an academic–industrial review of modelling in chemistry we at last put this book to bed and throw out, with some relief, the piled up, discarded drafts that were written on the way. We have enjoyed working together on *Chemistry through Models* and trust that others will in turn enjoy reading it, and find it useful.

As time progresses advances will naturally be made in the understanding of many of the subjects that we have used as examples but the principles underlying the modelling studies should remain valid even though a revaluation and re-interpretation of the data may become appropriate.

We should like to thank those who have discussed the book with us as it has developed, and in particular Dr W. Lawrie for his constructive criticism of the entire work, and Professor F. R. Bradbury. Our thanks are also due to Dr Colin Davies and Dr David Gosling for contributions to chapters 7 and 8 respectively, and finally to June Adam, Dorothy Grubb, Doreen Stammers and Barbara Stewart for their patience and diligence in typing the manuscripts.

Colin Suckling
Charles Suckling
Keith Suckling

1

AN INTRODUCTION TO MODELS

1.1 **What this book is about and why it was written**

A cheese paté bitten in mistake for a sweetmeat is likely to produce revulsion whatever the quality of the cheese and so, as this book is an unusual confection, it may be as well to give the reader at the outset some idea of its ingredients and why it has been thought worthwhile taking the trouble to concoct it.

Briefly, we have attempted to provide an introduction to modelling in general and a more advanced treatment of some important aspects of the use of models in academic and industrial fields ranging from speculative research to the developing of useful new products. The treatment seeks to bring out themes that are relevant to many of the diverse fields in which a chemist may nowadays be asked to contribute. Additionally, the central chapters of the book that deal with chemistry *per se*, provide a novel revision course for a substantial part of the subject.

The authors, two of whom were working in universities and one in industry, talking together about their work noticed that the discussion frequently turned to models and modelling, and that within these fields there were concepts which fitted together to make a framework in which their very different problems inter-related. The topics that happened to be under discussion at the time were, in the academic camps the use of model compounds in the study of biosynthetic pathways and of biological mechanisms, and on the industrial side the use of mathematical models of processes that were under development and of small-scale (semi-technical) plants as models of proposed manufacturing units. The authors were aware of the use of models in many other activities in which scientists participate and knew that these also would fit into the same framework of ideas.

There is, of course, nothing new in the observation that general concepts exist that are applicable to modelling across a wide area of pure and applied science, or in the recognition that these may be both intellectually stimulating and useful in practice, but though there are many excellent texts that have relevance for one or more aspects of modelling in chemistry, we have found neither a critical review of the practice of modelling in academic chemistry, though much work based on models has been published, nor a review of the principles of modelling as it may be applied over the many fields of activity in which a chemist may be required to practise at some stage in his career.

Passing reference to some important aspects of modelling, to the advantages and to the dangers, are to be found in the chemical literature. For example, Mislow (ref. 1) draws attention to a most important caveat that models should not be used under conditions for which they were not designed. Writing on the use of physical models of molecules he emphasises the fact that molecular models are built to reflect the behaviour of an 'average' molecule at near room temperature and cannot reflect the changes in the properties of molecular aggregates which result from variation in temperature. But, despite scattered allusions to some important aspects of modelling, there appears to be no general review.

On the other hand, one may discover equally readily papers that demonstrate the need for a clearer appreciation of modelling. The following quotation from a paper on flavin dependent oxido-reduction (ref. 2) while making, at least implicitly, some important points affecting the use of models, for example the need to relate them to their environment, suggests some difficulty in handling concepts in modelling.

> Models should not be any longer conceived as 'homunculi' or as 'bigger elephants than nature' or as micromolecular substitutes for macromolecular catalysts, but simply as derivatives of a coenzyme or a prosthetic group or even a mere protein functional group, which help in understanding the activation of the cofactor by the protein in terms of molecular structure.

But every vacant niche in the literature is not necessarily ripe for colonisation however attractive it may seem to potential authors. We would not have set to work on this book had we not believed that chemists who cannot use models effectively are likely to be seriously disadvantaged in the wider use of their skills.

The principal role of chemistry, it is often argued, is increasingly that of an enabling science. Certainly in the academic field the chemist's work overlaps more and more that of other disciplines, whoever may be said to take the lead. Beyond the university, chemists are applying themselves to an ever greater extent to tasks, often interdisciplinary, that lie outside the ambit of their own expertise. All this inevitably creates a need for chemists who can understand and interact effectively with practitioners of other disciplines, among whom must be included economists and sociologists. Indeed the increasing expectation that scientists will accept a share of the responsibility for the consequences of their work demands still wider skills, skills related to those of the designer.

Traditionally however the chemist has not been trained to cope with these requirements. Chemistry (like anatomy?) tempts its disciples to the relatively unstructured accumulation of facts. In academic courses, introductions to other disciplines are often introductions to facts rather than to underlying patterns of thought, which is to say to models. We are, of course, well aware of the great advances that have been made in teaching, in many centres, over the last decade or so; advances to which we hope our own presentation will prove to be complementary.

The scientist's social responsibility is, in large part though not totally, that of ensuring that the artifacts that he helps to create are not at odds with their working environment but rather that they are a good fit. The failures of the past have often occurred through concentrating on immediate technical problems to the exclusion of other factors. Too often the scientist perceives his contribution solely in terms of discovery or invention and fails to recognise that the translation of a new capability into practice – that is to say the process of innovation – requires or imposes changes in people's behaviour, changes that may offer opportunities to some, but for others impose constraints. The avoidance of mismatch between the products of innovation and their task environment is a problem of design, and the process of design employs models in many ways that extend from defining the task to testing the solutions.

One objective of this book is to suggest that inter-disciplinary interaction is most likely to be effective when it is based on an understanding of the conceptual models that the various disciplines use, rather than on mere learning of data, and also that useful models may be found in very unexpected places. In the hope of demonstrating this latter point we have not hesitated to quote rather than to paraphrase whenever we have felt that by doing so we can provide a direct contact with other people's models in a relevant context.

Much of this book is very relevant to disciplines other than chemistry and indeed to other industries, and examples are drawn from several areas, not only from those chemical. However, the unusual combination of topics that we present, though absolutely essential to the development of the theme we have chosen, undoubtedly creates difficulties. Some of these, for example the need for a cohesive treatment across all the chapters, are either met by the authors or not at all, but for others we must ask the indulgence of our readers. Those accustomed to modelling or, indeed, who are specialists in particular fields, may feel an unevenness in depth of treatment across the subject matter. This reflects partly the interest of the authors, but mainly the fact that for some of the topics with which we deal, for example process development, there is already an extensive literature compared with which the contribution in this book must be relatively light. On the other hand, though there are many published examples of modelling in biochemical synthesis, there is little in the way of analysis of its conceptual basis or its relationship to modelling in other fields. On this topic, therefore, our contribution breaks more new ground. Some of those who were good enough to read chapters in draft have suggested that the use of modelling in synthetic, degradative, and biological chemistry, and also in the discovery and development of new products, deserve books to themselves. If this be so, then perhaps the limited treatment that we have been able to give will serve as a stimulus for more extended studies.

It has also been suggested that since we cannot avoid trespassing, however reluctantly, on the preserves of the historians and philosophers of science, we should tramp these fields more boldly. But this is frankly not our purpose nor within our competence, though we do hope to encourage some readers who might not otherwise do so to explore these subjects for themselves. A beginning might be made with Mary B. Hesse's *Models and Analogies in Science* (ref. 3) in which a number of concepts that are of value in practical modelling are advanced during an exploration of the question of whether a theory must logically contain models and analogies if it is to be predictive and therefore accepted as scientific.

Modelling, like all techniques, can be a springboard or a straitjacket, creative or constricting. Moreover, the more powerful the tool the greater the damage when it is misused: the more fascinating the technique the more serious the risk of addiction. Our experience makes us all too alive to these possibilities and we take care to point out pitfalls insofar as we have been able to perceive them (chapter 2.7).

1.2 **What is a model?**

Modelling as considered in this book consists in constructing alternative, usually simpler forms of objects or concepts, in the expectation that the study of the model will shed light on the nature of those objects or concepts.

We need a name for that which is modelled and we shall use 'prototype' for this purpose. The *Shorter Oxford Dictionary* defines prototype as: 'the first or primary type of anything: a pattern, model, standard, exemplar, archetype'. Our use of prototype may slightly expand this connotation but there is at least one precedent in the publications of the Delft Hydraulics Laboratory. (Models are, of course, extensively used in investigating hydraulic problems, for example in studying the effects of tides and rivers on land conformation as well as in ship design. The Delft publications provide much useful commentary of general relevance to modelling, some of which we will mention in a later chapter.)

We will not use the word 'model' to connote an archetype or exemplar because, for our purposes, the essence of the model is that it is a restructuring or reformulation – that is to say a model is some transformation or other of its prototype.

It follows that a replica, in the sense of an exact copy, is not a model in our sense of word. Of course, what constitutes an 'exact copy' – a replica – depends on the depth and extent of the comparison. In a later chapter we refer to a radioactively labelled molecule as a model. It has been put to us that this is not a model of the unlabelled compound but a replica. If no radioactive effect were expected or sought then 'replica' might be an appropriate description – unless the label produced some other change that was significant for the purpose in hand. But this may well be the case. Isotope effects on reaction rates are common. As is well known, C–D and C–T bonds break roughly 7 and 14 times slower respectively than C–H. $^{12}C-^{14}C$ breaks in many reactions about 10% more slowly than $^{12}C-^{12}C$ and since ^{14}C may be present at a level of only 0.1%, this may have a significant effect on where and when to look for labelled products.

These are more recondite examples of differences in behaviour between labelled and unlabelled molecules that, being most unexpected, emphasise the need for constant care. One example must suffice. Some terpenes can be separated by differential absorbtion on a column that is composed of silica gel impregnated with silver nitrate, in which the C–C double bonds complex with the silver cations. In some instances when this technique has been used with terpenes labelled with tritium, the tritiated compound has been separated from the unlabelled molecules on the column. Failure to spot this could clearly have led to erroneous conclusions.

It is always safer, in experiment and in design, to think in terms of models rather than of replicas, and therefore to expect the unexpected discrepancy. Replica is a courtesy title that events may revoke at any time.

The identification of 'model' with 'prototype' as something to serve as a standard or to be copied is common usage. Kuhn uses the word in this sense when discussing those networks of scientific achievement that scientific communities acknowledge as providing, for the time being, the foundations for its further practice and which he has called paradigms: he writes (ref. 4)

> Achievements that share these two characteristics...
> (attracting an enduring group of adherents away from
> competing modes of scientific activity and being sufficiently
> open-ended to leave all sorts of problems to be solved)...
> I shall henceforth refer to as 'paradigms'...By choosing
> it (paradigm), I mean to suggest that some accepted
> examples of actual scientific practice – examples which
> include law, theory, application and instrumentation
> together provide models from which spring particular
> coherent traditions of scientific research.

Kuhn's paradigms are models in our sense also.

The different meanings of 'model' though potentially confusing should not cause difficulty as they are readily distinguishable. It is advisable, however, whenever the term model is encountered, to check from the context what is intended.

To talk of 'a model of a prototype' is not necessarily to imply that the prototype exists. A model may be constructed before its prototype as a partial realisation of something that one may wish to create. A map of a stretch of countryside is obviously a model of an existing prototype but, on the other hand, the ground plan of a proposed motorway is a model of a prototype that exists as yet only as an unrealised concept. The model is, in such cases, a tool to be used by the design team both to test the appropriateness of their plans and to assist in translating the final design into actuality.

Often, in the process of design, a series of models will be constructed each of which is intended to provide information on a particular aspect of the behaviour of the prototype. Sometimes the models will, as it were, converge on the prototype so that eventually one is accepted as defining and expressing for the time being *the* prototype, a standard and to be copied – as in the aircraft industry. Much of the expense of aircraft development arises from the fact that the reliability of so complex a system, consisting of thousands of interacting parts, cannot be confidently predicted from the behaviour of models each of which necessarily suppresses some interactions that may prove to be operationally important. As a result, extensive testing has to be done at full scale and with the entire aircraft. Likewise in the design of chemical plant it is sometimes necessary to undertake tests at full scale, though the overall problem is less because it is usually easier to provide protection against the consequences of component failure.

1.3 Types of model

Depending on the purpose in view one can define many different types of model and classify in various ways. A useful categorisation of models and a formal analysis of their taxonomy and development will be found in Mihram's paper *The Modelling Process* (ref. 5).

For the purpose of this introductory chapter it will be sufficient to classify models broadly as physical or conceptual, the latter including mathematical models. That models of these types are fundamental to learning – in organising ideas in such a way as to be able to predict effectively from experience – is very clearly explained in Bruner's *Towards a Theory of Instruction* (ref. 6) which book, incidentally, we would strongly recommend to anyone who is interested in inter-personal relationships (and who should not be?).

To go beyond the responding to information encountered on a single occasion and to predict the future on the basis of experience, the individual must store information in some structured way that corresponds to his environment. Bruner considers this to be the essential process of learning and suggests that the child develops three types of model.

The first way in which the individual can translate his experience into a model of the world is, Bruner suggests, through action.

> We know many things for which we have no imagery and no words, and they are very hard to teach to anybody by the use of either words, or diagrams and pictures. If you have tried to coach somebody at tennis or ski-ing or to teach a child to ride a bike, you would have been struck by the wordlessness and the diagrammatic impotence of the teaching process. [Ref. 6*a*.]

This type of modelling Bruner calls enactive, and we can see a close parallel in the behaviour of those scientists and engineers who like to begin tackling a problem by experiment, by doing something with their hands, by twiddling the knobs.

Bruner's second system is based on summarising images and he calls it iconic. Some models are, in one way or another, physical representations of their prototype. Examples are: the small-scale plant (often called semi-technical or pilot plant) that is built to provide data for the design of full-scale manufacturing plant, molecular models, maps (including for example plots of electron density and of X-ray diffraction data), and model compounds used in synthetic, degradative and biological studies. It is this kind of model that is called iconic and which Ackoff (ref. 7) defines as follows:

> Iconic models are large or small scale representations of states, objects, or events. Because they represent the relevant properties of the real thing by those properties themselves, with only a transformation in scale, iconic models *look like* what they represent. For example, road maps and aerial photographs represent distances between and relative positions of places and routes between them. With respect to these relevant properties such maps or photographs look like the real thing: they differ from it with respect to these properties only in scale. Flow charts which show the processing of material or information may also be iconic models, as may be floor plans or other types of diagram.

This definition gives an indication of the difficulty in attempting a logical and consistent classification of models. How, for example, could one define 'look like' to permit it to cover all examples given in the quotation: and does a flow chart differ from its prototype only in respect of scale?

The third type of model, the symbolic or conceptual, is essentially a mental construct that may range from the simply descriptive to the rigorously analytical and in which the symbolism may be as varied and loosely defined as a pattern of thought or as precisely defined as an algebraic equation. Like other types of model, symbolic models may be static or steady-state models or they may be dynamic and represent changes that occur with the passage of time.

Scientists are often dissatisfied with models that are not developed beyond the descriptive stage since at that level no quantification and therefore no rigorous experimental testing is possible. Nevertheless, much that is vital in modelling takes place at the descriptive stage. Experienced workers will agree that more models fail to be useful, or even positively mislead because they are inappropriate than fail through neglect of rigour or through inadequacy in mathematical techniques. If we fail to ensure that our model is relevant to the problem in hand, that significant characteristics of the prototype and of its relationship to its working environment have not been over-looked, then no degree of sophistication in its representation or mani-pulation can remedy the deficiency. It is at the descriptive stage that most decisions on these important matters are taken and it follows that a good descriptive model may, indeed should, present a useful analysis of the factors that are significant in the situation under study.

Perhaps the most important of his conceptual models is each person's private representation of the world about him. Beer, in his book *Decision and Control* (ref. 8) which is strongly recommended as pro-vocative, stimulating and useful (as is also his latest book *Platform for Change* (ref. 9)) writes

> Rational conduct depends not only upon knowing what is
> really happening and being able to interpret it, but on
> having present in our minds a *representation* of what is
> going to happen next. This representation is not an account
> of what is the case, but a continuous prognosis of what is
> about to be reported to us as being the case. It is a
> prognosis continuously corrected by the feedback. Let us
> call this mental representation of the world that is not a
> direct perception of the world a model of the world.

This model is, of course, not fully verbalised, indeed it may not be entirely at the conscious level. It is a dynamic model that predicts the behaviour of the real world. Expectations are compared with what actually happens and the model is continuously updated to eliminate discrepancies.

In the absence of feedback, as seems to be the case in dreaming, bizarre results may ensue. In the waking state, when apparently important messages reach the consciousness – or perhaps before – we attempt to verify them. On hearing the fire alarm we look for flames or smoke and sniff the air. The result of these investigations constitutes an element of feedback into the system. In that state of sleep in which intense dreams occur we are unable to interrogate the environment and are deprived of recourse to our enactive models. All but the most intense external stimuli are shut out. The patterns of which we are conscious, though they may be entirely unrelated to the current state of the real world or, sometimes, an attempted rationalisation of one intense stimulus (ref. 10) are accepted as a model of it and may stimulate emotion as intense as that which we experience when awake. Dream experiences, at least frame by frame, may be heavily transformed models of past experience but the correspondences are often difficult to identify.

Bruner sums up his thinking on the three types of model thus (ref. 6b)

> What comes out of this picture is a view of human beings who have developed three parallel systems for processing information and representing it – one through manipulation and action, one through perceptual organisation and imagery, and one through symbolic apparatus. It's not that these are 'stages' in any sense; they are rather emphasis in development. You must get the perceptual field organised around your own person as centre before you can impose other, less egocentric axes upon it, for example. In the end, a mature organism seems to have gone through a process of elaborating three systems of skill that correspond to the three major tool systems to which he must link himself for full expression of his capacities, tools for the hand, for the distance receptors, and for the process of reflection.

We have quoted Bruner at some length not merely because he provides a concise and elegant psychological basis for the two major categories of model: iconic, and symbolic or conceptual, and in his enactive model for the practical scientist's often irresistible temptation to experiment on the instant, but also because he exposes very clearly an unconscious process which is obviously of enormous potential also at the conscious level, at which however it too rarely achieves expression.

Though models, as Bruner suggests, may be fundamental elements in learning, most people first encounter the words 'model' and 'modelling' through toys, which are obviously iconic models. As a result, a few find the extension of the idea of modelling to conceptual patterns to be strange and difficult. Nevertheless, a brief consideration of simple toy models will introduce some important matters that are fundamental to the use of all types of model and with which we shall deal at some length later in the book.

Models are used as partial substitutes for their prototypes – to assist in designing, understanding, predicting the behaviour of, controlling, or experiencing emotions associated with the prototype. In order to perform these functions they must represent some significant characteristics of their prototype and these characteristics may be a matter of appearance, of operation, or both. Which and how many properties are significant in a particular case and how they are to be represented in the model is at the heart of the activity of modelling and these questions will receive much attention in subsequent pages. For the moment we note that the answers will depend very much on the purpose to which the model is to be put.

You can have a lot of fun playing boats with any old piece of wood and, with advantage, eke out its imperfections with your thoughts, but if you buy a model boat you will expect it to show at least the most obvious characteristics that are associated with the prototype 'boat'. Operationally the piece of wood preserves the essential characteristic – it floats – but it may be a bit inadequate in appearance. Even young children expect at least a sharp and a blunt end and perhaps some sort of superstructure. The more sophisticated customer may demand a lot more detail and perhaps pay a lot more for a model 'to scale'. A model boat might be made for the fun and the exercise of skill in the making, or to keep in a glass case, or to sail, or as a step in designing a new boat or for many other reasons. Some of the characteristics of the prototype, for example the ship's lines, may be as important to the modeller with purely aesthetic intentions as they are to a ship's designer, but there will be many characteristics that the former will regard as important but which the latter will be able to ignore in his model, and vice versa.

The need to suit the model to its purpose is elementary but, like much that is elementary, it is very important and will reappear in the discussion in subsequent chapters. But who defines the purpose? The modeller may, like the designer, the inventor and the artist, sometimes work for himself but sometimes for a client. The boat modeller's clients range from the child to someone buying an ocean cruiser. To the student modelling chemical reactions in the laboratory his professor may be the client. In industry the clients calling for a model in the form of a semi-technical plant may include those who have to design the full scale plant – who want design data, and those who have to vote the money for it – who want to be convinced by successful semi-technical runs that the full-scale plant will be both safe and economically sound. So the modeller must have regard not solely to his own purposes and objectives but also to those of his clients, not only because these purposes define what the model is required to do but also because a skillfully designed and constructed model can be a powerful means of communication and as such help to establish understanding between the modeller and those with whom he must collaborate.

1.4 Models and problem recognition

In comparison with problem solving, for which a plethora of techniques (including many types of model (refs. 8, 11, and 12) have been devised, problem recognition has been neglected. Franks in his very useful book *Modelling and Simulation in Chemical Engineering* (ref. 13) writes

> The first step, perhaps the most important one, is that of problem definition, yet it is not possible to establish rules for problem definition that are sufficiently general to be useful. Technical problems are so diverse that it is up to the analyst to state clearly the nature of the individual problem. This will establish a definite objective for the analysis and is invaluable in outlining a path from the problem to the solution.

We can therefore usefully illustrate modelling from the field of problem recognition and definition which, moreover, is an important element in designing.

A problem exists when a threat or opportunity is recognised and one is not sure how best to act in order to counter the one or take advantage of the other. Problem recognition is clearly a matter of no small importance in which insensitivity may lead at best to missed opportunity and at worst to disaster.

Problems do not usually present themselves in clearly delineated form and often involve complexities of uncertainty, cost, benefit, allocation of scarce resources, and so on. These points are clearly brought out in the following quotation from Pounds (ref. 14). Although it refers to managers and is primarily directed to a business situation, it should be seen as highly relevant to decision taking in academic environments.

> But the manager's job is not only to solve well-defined problems. He must somehow identify the problems to be solved. He must somehow assess the cost of analysis and its potential return. He must allocate resources to questions before he knows their answers. To many managers and students of management the availability of formal problem solving procedures serves only to highlight those parts of a manager's job with which these procedures do *not* deal – problem identification, the assignment of problem priority, and the allocation of scarce resources to problems. These tasks which must be performed without the benefit of a well-defined body of theory may be among the most critical of the manager's decision making responsibilities.

The use of models for problem identification has been developed by Loasby (ref. 15a). His work was stimulated by dissatisfaction with the model of decision making in firms that is commonly used by economists, that is to say the marginalist approach which suggests that businessmen adjust their policies to take advantage of every slight shift in balance of opportunities by making marginal adjustments to meet marginal changes in circumstances. That is not of course to say that the making of opportunistic adjustments is not of fundamental importance in the management of a business, it is very important. Loasby remarks that if economists use the wrong model then they will come to irrelevant conclusions and give bad advice.

The chemical engineer will recognise the difficulty inherent in designing a control system to maintain a closely defined track under varying conditions, the need for a control system that does not simply apply corrections proportional to the divergence of actual value from desired value but adjusts the correction according to the rate of departure, that is to say by derivative action and also according to the integrated discrepancy over time from the desired value. In fact the elaborate monitoring system that would be necessary to provide the data to enable sensitive marginalist adjustment to be made in a business would be far too expensive even if it were possible. Moreover, because of human limitations and also limitations in the design of suitable computer programmes – whatever computer experts may say to the contrary – it is quite implausible.

Before going on to describe some models that can be used in problem identification Loasby points out that four conditions need to be satisfied before action is initiated. First, problem recognition itself, that is to say awareness that some threat or opportunity exists; second, the acceptance by some person, or group, of responsibility for dealing with it; third, a sufficient incentive to cause those accepting responsibility to act; and fourth, a belief by those accepting responsibility that the possibility of finding a solution exists.

Loasby then analyses the process of becoming aware of a threat or opportunity in terms of a comparing of models, one of the situation as it is and the second of the situation as it might be. Problems are defined by difference – a procedure closely related to the operation of feedback in Beer's model (ref. 8a).

What kinds of model are there with which to make a comparison, models that may show up unsatisfactory features in the actual state of things? Among those proposed by Loasby are models of competitive performance, historical, planning and imaginative models. Comparison with competitors is an obvious way of assessing one's own perform- ance. Care must be taken to ensure that like is compared with like. One may be misled if accounting conventions differ: for example, one firm may include in research and development costs the work in the factory designed to improve process performance quickly or perhaps technical support for sales, while the other leaves both out.

Historical models can be used in two ways. One can either compare the model of a favourable historical situation with that of the present to see what, if anything, is lacking now, or find a time in the past which matches the present in many significant respects, construct a dynamic model that predicts what did in fact happen, then with a similar dynamic model of the current situation operate it in the same way and examine what it predicts. That most frequent resource of the re- searcher, the 'off-the-peg' model constructed from the literature which usually provides helpful analogies and which is discussed in chapter 5, may be regarded as a historical model though it is used more to solve problems than to define them. Reliance on historical models presupposes that the past will prove, in some important ways, a guide to the future; that the environment with which the historical model interacts is the same now and in the near future as it was then. Imaginative study of what has been assumed to remain constant may be required as a corrective.

Though the dangers of unthinking extrapolation in time are great there are situations which, if they recur at all, are likely to contain many predictable factors. If these situations pose problems it may be better to devise in advance strategies to deal with them rather than to defer the thinking until things go wrong and pressure mounts. The use of check lists for indicating steps to be taken in, say, designing a plant (chapter 7) or developing a new product (chapter 8) are based on this concept. The drills for flying an aeroplane are based on the expectation of a high degree of predictability. Most probabilities and indeed improbabilities can be foreseen if adequate time and imagination are expended. A historical model of normality and a partly historical and partly imaginative model of possible aberrations and their consequences are worked out. The pilot's model of normality is a given pattern of instrument readings, deviations from this pattern alert him to a discrepancy between models, though he may not think in these terms.

Does the pilot confront reality and not merely a model when he faces his rows of instrument dials, gauges and warning lights? If a tree tilts acutely we expect its fall to be imminent. If we affix a tilt meter to it and read that the angle of tilt is $57°$ are we not still in contact with reality and not looking at a model of the tree's behaviour? The two situations differ in the important respect that the tilt meter merely quantifies the position of the tree that we observe directly, and if the tilt meter is faulty we will not believe it. The pilot, on the other hand, is often unable to perceive directly the attitude of his aeroplane and is dependent on the veracity of his instruments, the readings of which he usually cannot verify by direct observation. Of course, the pilot's training is designed to enable him to respond to and rely on his instruments *as if they were* the real world. But they are not, he cannot look inside the fuel tanks to see how much he has left. The safety of the crew and passengers depends firstly on the skill of the designers in identifying a pattern of information that is so good and reliable a model of the aeroplane and its relevant environment that the pilot can accept it as a substitute for 'the real world', secondly on the pilot's skill and alertness, and also, of course, on the skill and care of those who make and maintain the equipment. The model must be adequate, relevant, explicit, robust, and well maintained.

One of us once asked a 747 pilot what he did if anything went wrong – his answer – sit on my hands and think what is the drill. The situation contrasts with that in some plants in the chemical industry where trouble tends to be combated by *ad hoc* analysis. There is room for a lot more explicit formulation in advance of models of aberrant behaviour, and definition of emergency drills in advance.

In contrast to historical models, planning models and imaginative models both, as Loasby puts it (ref. 15*b*), 'cross the bridge from past to future'. But while planning models prescribe targets (what should be), imaginative models suggest hypotheses (what might be) and, while the process leading to the planning model can be analysed, that leading to an imaginative model involves a 'creative leap' which can at best be rationalised.

A compelling example of the reference model as a practical tool in an important field of problem identification – medical diagnosis – is to be found in an excellent guide to the techniques of study that the Department of Clinical Anatomy in the University of Liverpool provided for its students. A repeatedly emphasised theme is that the physician must have clearly in his mind a picture of the range of variability that lies within the bounds of normality, because this will provide him with a basis for comparison and permit the early recognition of signs of trouble. In other words, the student must piece together for himself a model (though they do not use this term) of the normal and the guide suggests that this objective deserves a more positive approach than it often receives.

> Individual variations in the anatomical, physical and mental make-up of a normal healthy person are so great that a thorough grounding in the normal, and in the normal variations which occur, is an essential preliminary to the study of disease. Many common conditions consist of variations from normal and hardly amount to disease...
> The responsibility of the clinical teacher is to interest the student in the normal, the common rather than the unusual, the more important rather than the rare, in persons and people rather than in cases, in health as well as disease.
> He must instil into the student a thorough knowledge of the normal as an essential basis for the study of the abnormal...

1.5 **The gas law as a model**

Theories, hypotheses and laws all fall within the scope of the definition of model that has been adopted in this book. A look at one of the elementary laws of physics will draw our attention to some considerations that are of general importance in modelling.

The simple form of the general gas law expresses the relationship between the pressure p, the volume v, the temperature T, and the quantity n of gas in the algebraic expression $pv = nRT$, where R is a constant. The derivation of this deterministic equation from the statistically based kinetic theory of gases, which describes the rapid, random motion of gas molecules is straightforward. This expression is a mathematical model of the readings of an appropriate set of instruments attached to a closed volume of gas. If the readings of any three of these instruments are substituted for the corresponding variables in the model, thus becoming the input to the model, then the value of the fourth variable that is demanded by the identity will, of course, correspond to the reading of the fourth instrument and constitute the output from the model.

The equation is referred to as the ideal gas law. The word 'ideal' immediately suggests a departure from reality, an abstraction. As is well known, the ideal gas law does not model precisely the behaviour of gases because it neglects some relevant factors – for example, interactions between gas molecules and also the volume of the molecules themselves. In many applications these factors may indeed be neglected with impunity but sometimes they may not, and when using models it is essential to keep the proven range of applicability well in mind.

A model may be misleading if an essential characteristic of the prototype has not been represented. But it is not sufficient merely to include all the variables, the limits within which they are free to move must also be expressed. For example the gas law as formulated above does not hold for all values of the variables: negative values are excluded, p must be less than the critical pressure and T must be less than the temperature at which the gas dissociates significantly. Moreover, the equation relates only to a closed system. These limitations can, of course, be specified as part of the model if we know the value of the critical pressure and the relationship between dissociation and temperature. A need to keep clearly in mind the inbuilt limitations is of great importance in all conceptual models.

If the 'ideal' model does not meet the needs of the situation then it must be amplified in some way or another. Preferably any new parameter that is introduced should have a physical significance or interpretation, but this is not always possible and resort has sometimes to be made to empirical adjustments. The relative merits of empirically and theoretically based models will be discussed in chapter 2. To improve the gas law both approaches have been tried. By far the most successful equation is that proposed by van der Waals

$$(p+a/v^2)(v-b) = RT \quad (a \text{ and } b \text{ are constants}).$$

The added terms can be understood as representing the state of affairs near a wall of the vessel containing the gas, where account must be taken of the change in pressure which results from the fact that molecules near the walls are not uniformly surrounded by other gas molecules. Thus a/v^2 represents the attraction of molecules near the walls for each other and b takes account of the volume of the molecule. If one determines b for a gas, it is possible to estimate the diameter of a molecule of that gas, and so the van der Waals' equation embodies a physical meaning in its constants.

1.6 **Environments**

Some models are made for the fun of it or just for looking at, valid reasons indeed for modelling but not our principal concern in this book. Most of the models that we consider are constructed for the help that they may give in understanding their prototypes; in designing or in explaining, in predicting or controlling its behaviour. For practical reasons it is necessary to limit the scope of the modelling, to split out as it were, the prototype from the rest of the universe. Nevertheless, it is the behaviour of the prototype in its place in the real world, that is to say in its environment, that is the ultimate concern of the modeller. If the modeller focusses his entire attention on the prototype and neglects to visualise how it will look in its task environment he is likely to run into trouble – as is especially obvious when a model is used in a design project. The interaction between prototype and environment is two way. If you introduce an alien species you must ask not only whether it will survive but also what it may exterminate. Not only will the environment exert its pressure on the prototype, enriching its potentiality but also imposing constraints, but the prototype simultaneously will alter the environment, sometimes in ways that are of little significance, but often otherwise.

As an example, let us consider some aspects of the design of a chemical plant, a subject that will be dealt with more fully in chapter 7. The potential impact of the operating unit on the physical environment – the world we live in – in terms of its hazards, effluents, noise and so on is of obvious importance and a semi-technical (ST) plant built to provide design data will be required to furnish information on these matters. Indeed the minimising of hazards and containment of effluent will be a fundamental factor throughout the development of the process from the laboratory onwards. If, as has often been the case in the past, the problems of objectionable wastes or of other nuisances are given inadequate consideration, then this defect in the modelling process may lead to serious trouble, and so it is clear that an efficacious model will have to relate to the ultimate environment of the prototype.

Can we not expect, however, that the operation of an ST plant will automatically highlight potential environmental problems? Not necessarily. They may, for instance, be obscured by the scale factor. A tiny compressor may run quietly though the full-scale machine shriek, a trickle of effluent may be easily disposed of without polluting, being so small in quantity, while giving no indication that a serious problem will exist at full scale. This possibility has to be given serious and specific thought – it cannot be left to chance discovery. One cannot ignore the real life environment when planning a modelling exercise, it must be taken into account in the design both of the model and of the experimental programme that will be undertaken with it.

No mention has so far been made of any impact that the environment – in this case a particular factory and its surroundings – may have on the new plant. This is usually very important, for example, it is often the case that a plant forms one link in a chain of chemical processes so that the product from one becomes a raw material for the next. It is all too easy to operate a laboratory or ST model of a new process using raw materials that differ significantly from those that will be produced by plants in the factory. Minor constituents in a feedstock can have a major effect on a process, as has been frequently observed in processes that use catalysts, the activities and specificities of which can, as is well known, often be altered by very small quantities of 'poisons' or 'promoters'.

In thinking in terms of prototype and environment we focus attention on the structure of the total system that we are studying and on the danger of failing to take into account important interactions between prototype and the world external to it. The prototype itself will also often have a complex structure, not all elements of which will be represented in the model. The problem of deciding which interactions within the prototype to include and which to leave out is essentially the same as that which arises in relating the prototype to its external environment, and it has proved helpful to consider the components of the prototype in their entirety as its *internal* environment.

The concept of internal environment is, of course, relevant to semi-technical plants. It may be necessary to recover and re-cycle unconverted raw materials. However, the recovered stream may well contain traces of materials produced in the process and not present in the original feedstock: these may have a profound effect which would be missed if the model omitted the re-cycle. Yet the inclusion of a separation and re-cycle stage may add substantially to the ST cost. Likewise, the particular geometry of the plant may lead to local inhomogeneities. For example, a film of moisture may build up at a particular cold spot in, say, a distillation column which could lead to serious local corrosion. Factors such as these appertaining to the internal environment of the intended prototype obviously need to be taken into account when designing and operating the model. Risk can usually be reduced at the cost of further and more elaborate experimentation, but how far to go is a matter of judgment.

The concept of internal and external environments is developed by Herbert Simon in his very stimulating book *The Sciences of the Artificial*. In the following quotation, which provides a very apposite summary of the concept, Simon writes of the functioning of man-made artifacts in their environment, that is to say of design (ref. 16).

> Fulfilment of purpose or adaptation to a goal involves a relation among three terms: the purpose or goal, the character of the artifact, and the environment in which the artifact performs. When we think of a clock, for example, in terms of purpose, we may use the child's definition: 'a clock is to tell time'. When we focus our attention on the clock itself, we may describe it in terms of arrangements of gears, and of the application of the forces of springs or gravity on a weight or pendulum.
>
> But we may also consider clocks in relation to the environment in which they are to be used. Sundials perform as clocks *in sunny climates* – they are more useful in

Phoenix than in Boston, and of no use at all during the Arctic winter. Devising a clock that would tell time, on a rolling and pitching ship, with sufficient accuracy to determine longitude was one of the great adventures of eighteenth-century science and technology. To perform in this difficult environment, the clock had to be endowed with many delicate properties, some of them largely or totally irrelevant to the performance of a landlubber's clock.

Natural science impinges on an artifact through two of the three terms of the relation that characterises it: the structure of the artifact itself, and the environment in which it performs. Whether a clock will in fact tell time depends on its internal construction and where it is placed. Whether a knife will cut depends on the material of its blade and the hardness of the substance to which it is applied.

We can view the matter quite symmetrically. An artifact can be thought of as a meeting point – an 'interface' in today's terms – between an 'inner' environment, the substance and organisation of the artifact itself, and an 'outer' environment, the surrounds in which it operates. If the inner environment is appropriate to the outer environment, or vice versa, the artifact will serve its intended purpose . . .

This way of viewing artifacts applies equally well to the many things that are not man-made – to all things, in fact, that can be regarded as 'adapted' to some situation; and in particular, it applies to the living systems that have evolved through the forces of organic evolution.

So far we have considered the internal and external environment of the prototype in relation to the behaviour of the prototype on which the model is to throw light. We must, however, go further and consider the model in its environment. A model ship when being used as part of the design process to furnish hydrodynamic data is obviously not to be tested in the open sea but rather in a specially designed tank in which the flow of currents and the size of waves can be scaled to match the scale of the ship – in other words, there has to be a model of the external environment.

Occasionally but rarely, the external environment is of negligible importance in modelling. A model ship constructed purely for display does not need to relate in any physical particular to the sea, but even so an imagined relationship is essential if the model is to be more than a form in itself.

As we shall also see later, it is important to recognise the range of prototype environment to which the deductions drawn from the model will be appropriate: a robust model will be relevant to the prototype behaviour even when the conditions under which the prototype finds itself are varied widely, whereas a weak model will be significant only within narrowly defined limits.

1.7 Why model?

Though modelling is, at least implicitly, an inescapable if subconscious component of a scientist's pattern of thought, this book is essentially concerned with explicit uses of modelling techniques. Why should we resort, deliberately, to modelling? One may model, as it were reluctantly, because it is too complicated, expensive, dangerous or inconvenient to work with the prototype. On the other hand there are situations in which despite the availability of the prototype for experimentation, one chooses to use a model.

We consider these two situations in turn:

(1) It is too complicated, expensive, dangerous or inconvenient to work with the prototype.

There are certain situations in which access to the prototype may be impossible. For example, it is not acceptable to use human beings for testing new drugs until everything possible has been done to ensure that they will be beneficial, although final evaluation can only be made in man. Consequently organisms of various kinds, be they cells, isolated organs, or whole animals, are used as models with the object of predicting the response of the human to the drug. Sometimes an inanimate system may be included in the screening. In chapter 8 we shall look more closely at procedures used in evaluating the potential of candidate drugs and other products, and suggest that the screens are, in effect, models of the environment in which the product in question has to be effective.

Coenzyme B_{12} (see chapter 5) is one of the structurally most complex non-polymeric natural molecules and a formal total synthesis of this molecule has been the target of a vast and costly research effort. Accordingly, in the development of the synthesis advantage was taken of model compounds. As if unsatisfied by presenting a prodigious synthetic problem, coenzyme B_{12} mediates a number of enzyme catalysed reactions which have some of the most puzzlingly complex mechanisms yet investigated. The present level of understanding of these reactions owes a good deal to the study of the properties of highly simplified models of the enzyme.

Time may be an insuperable barrier. Any situation that belongs to the past, for example that which obtained when life was developing on this planet, is obviously inaccessible and can be studied only through imaginative models. Looking to the future, many products such as building materials are worthless if not serviceable for many years. In this situation a model environment may have to be constructed in which stresses are magnified while the time scale is contracted, in an effort to predict long term behaviour from short term tests.

The impossible shades off into the inconvenient. It is possible to work out a complicated multi-stage synthesis with a radioactively labelled isomer, but this is often expensive because of the cost of isotopes and of safety procedures. Better then to blaze the trail with the inactive prototype which will nevertheless be essentially identical in chemical behaviour. A possible strategy in the development of new plant for chemical manufacture is to build a manufacturing unit on the basis of laboratory data and then work out the snags with the full-scale ironware. This can sometimes be quick but is almost always expensive; it may also be highly dangerous, and a model in the form of a semi-technical plant is usually considered to be an essential step in the design process.

(2) There are on the other hand situations in which one may decide to model even though the possibility of working directly with the prototype exists. The following are some of the reasons for proceeding in this way.

Complex situations are often difficult to grasp, difficult to describe and difficult to discuss. Their elusive nature may often be grappled with by constructing a model that serves both as a description of the situation and provides a framework for discussion. This is especially important in complex projects which demand contributions from many disciplines. Such a model is useful not only in representing a static situation but also and especially in coping with instability in the problem: with the inevitable changes in objective and opportunity that occur in the course of any significant creative exercise when there is need of a clearly expressed record, a clearly expressed problem description which can be updated as necessary.

Models can also be very powerful in the exploration of alternative policies, a subject that is explored by Davies and his co-authors in the book *The Humane Technologist* (ref. 17).

In a book that should be of considerable value to research workers as well as designers, J. Christopher Jones writes (ref. 12*a*).

> Without something equivalent to a drawing (in which to store, and to manipulate relationships . . .) the system designer is not free to concentrate upon one bit of the problem at a time and he has no medium in which to communicate the essence of the mental imagery with which he could conceive a tentative solution which would enable him to drastically shorten his search

and he points out (ref. 12*b*) that one purpose of a scale drawing, the main instrument of the traditional designer, is to give him a much greater 'perceptual span' than was available to the craftsman. It gives him the freedom to alter the shape of the product as a whole, instead of being tied, as the craftsman is, to making minor changes.

A complex situation in model form can be studied piece by piece while at the same time the interaction of the system as a whole need not be lost. This possibility is general but it is especially obvious in mathematical models formulated for the computer when, not only is there a facility for studying a variety of configurations of the model and a variety of values of parameters within them, but as a necessary means of programming the computer a language has been devised which enables one to structure complex problems as interlocking sub-models within the total programme which constitutes the master model.

We humans have very constricted short term memories. It has been argued (ref. 18) that only about seven basic units of information can be kept available at any one time in the short term memory – and that about five seconds are required to transfer an item to the long term store.

Since most problems require the structuring and restructuring of complex patterns a formal, succinct, written symbolism to take the load from the short term memory is invaluable. This is provided, most obviously, for models that can be expressed in algebraic terms, by computer languages.

The importance of the model as a representation that can provide a communicable special vocabulary and syntax relevant to the problem in question cannot be overstressed. To quote Bruner once more (ref. 6*c*).

> Language is perhaps the ideal example of . . . a . . . powerful technology with its power not only for communication but for encoding 'reality', for representing matters remote as well as immediate, and for doing all these things according to rules that permit us both to represent 'reality' and to transform it by . . . appropriate rules,

and speaking of the great clarification of thinking that often comes when one reads one's first attempts at expressing an idea. Bruner lays great stress on the dialogue between the thinker and his written words

> In such reflections notation of one sort or another surely becomes enormously important, whether by models, pictures, words or mathematical symbols. And there is a gap and we know too little about the use of a notebook, the sketch, the outline in reflective work.

Finally, there are situations in which a model may serve to generate variety. The concept of variety, in this context, is borrowed from cybernetics. Useful discussions can be found in Beer (ref. 8) and Ashby (ref. 19). Of course, a model does not usually represent all the characteristics of its prototype; only those that are believed to be significant for the purpose in view are included and a simplification, which implies a loss in variety, necessarily ensues. Nevertheless, models even though they be simplifications may open up new possibilities in many ways.

For example a model as a tool in designing can suggest possibilities that had not previously been envisaged. This is obvious and does not need exemplifications. Moreover models lend themselves to experimentation. The question 'what would happen if circumstances were otherwise?' can be asked and, if the model is robust and realistic, often answered. Loss in certainty which is consequent on working with a model – an abstraction – rather than with real life – the prototype – is accompanied by gain in freedom to change the circumstances. Experimentation based on such possibilities may lead to new insights.

The use of molecules without isotopic labels as models in scouting experiments is mentioned earlier in this section. But conversely the labelled, isotope containing molecule may equally be regarded as a model of the 'natural' inactive molecule, a model which carries more information and can therefore generate more variety.

The possibilities of modelling as a creative technique are argued further in the next chapter.

MAKING MODELS

2.1 A modelling process

Our most basic learning task, that of acquiring the ability to organise information about the world into a usable structure, requires us to construct models. Bruner (ref. 1) has described very clearly the various types of model that children develop for this purpose spontaneously and unselfconsciously and his classification, to which we have already referred, into enactive (tools for the hand), iconic (tools for the distance receptors), and symbolic (tools for the processes of reflection) is highly relevant also to explicit model building. So Le Bourgeois Gentilhomme had not only been using prose for forty years without knowing it, he had also been modelling.

As Bruner makes clear, modelling is so deeply woven into the fabric of our thinking that, like walking, we do it without having to analyse or even to be aware of the mechanism of the process. Nevertheless it is useful to develop a structured approach to modelling, and we will use the following process.

(1) Recognise the existence of a problem and decide to tackle it.
(2) Delineate the system to be studied.
(3) Formulate questions to be asked.
(4) Construct the model.
(5) Run the model.
(6) Analyse the results and their implications.

2.2 Recognise the existence of a problem and decide to tackle it

When we want to do something – to discover, understand, learn, plan, design, or build – and our information or our analysis of the situation is not adequate for us to decide with reasonable confidence how to go about it, then we have a problem. Uncertainty in the face of need to act is the essence of a problem. Research is often the best tool for uncertainty reduction and, in the situations that we are studying, modelling is often the most appropriate research procedure.

Everyday words cannot be expropriated exclusively for techical uses and so, provided that the context excludes confusion, the word 'problem' will be used in this book in its everyday connotations as well as in the special usage that we are proposing.

The effective pursuit of any but the simplest, unvarying objective demands an adaptive strategy. As time passes and events unfold, threats and opportunities – some foreseen, some unexpected, some neither foreseen nor perceived – will arise; threats that endanger the achieving of objectives, and opportunities to improve on them. But so long as the threat or opportunity remains unnoticed, no problem will exist and there will consequently be no stimulus to act; no attempt to avert the danger or to take advantage of the opportunity.

Threats undetected often give rise to major problems. How can we scan the environment systematically in order to spot threats and opportunities in time to act economically and effectively? An important general procedure has already been discussed in the previous chapter where, following Pounds and Loasby, it was suggested that a model of the situation as it is should be compared with a reference model which might indicate in what respect things could be better.

If such a comparison can be made early enough, then corrective action may be obvious and no problem need arise. There are many stable ongoing systems, for example chemical plant and businesses, for which a set of measurable properties that define satisfactory performance can be identified and may be used as a reference model against which the performance of the system under surveillance can be monitored continuously. Many devices of interest to the chemist, ranging from simple laboratory equipment to complex control systems, fall into this category. They contain a reference model in which are embodied the limits within which control functions may be permitted to vary, an input from instruments which measure the actual level of these functions in the system that is being controlled, and devices for comparing the two values and making the necessary corrections to the operating variables.

Analogous monitoring systems, based on the collection and comparison of operating data such as sales information, examination marks, or product quality can be devised for alerting those responsible for other types of system to the need for action. In all cases realistic models and reliable realisations of them are essential. A control system that fails to respond when it should, or which sounds off when there is no discrepancy is worse than useless.

The extent to which we can devise systems that will protect us against unwanted surprises depends on how well our control models can be designed to take account of future events. Confidence that all potential causes of serious failure have been anticipated has often to be bought at high cost in terms of resources used and there is continuous debate on the question of how much research should be done on new processes, new products, new machines in order to reduce the risk of unsatisfactory performance to an acceptable level. Since there are more problems to be tackled than can be coped with by available researchers, the true cost of a piece of research is not always simply the expense incurred. It is rather the probable value of the best alternative project that is not undertaken. That is to say it is the value of the lost opportunity, the so-called opportunity cost.

The extent of the research that might reasonably be done to assist in taking a particular decision, by reducing uncertainty, will be related to the probability of deciding wrongly in the absence of further information, the penalty for being wrong, the probability that further information will improve the decision, the probable value of the expected improvement, and the cost of doing the research. A good deal of work has been done on the theoretical structuring of the sort of question raised in this paragraph, and on the possibility of finding ways of quantifying expectations. Anyone wishing to follow this might well begin with *Management Science. A Bayesian Introduction* by W. T. Morris (ref. 2). The earlier difficulties are recognised, the less will be the cost of dealing with them. The cost of switching the course of any complex activity whether it be building a new aeroplane or a chemical plant, working out a long synthesis or developing a teaching course or a new product is likely to escalate sharply as one goes farther down the track.

Some excuse for seeking more information can always be found but the best is often the enemy of the good. Delay in coming to a decision may not merely put off the enjoying of whatever benefit is expected but can often frustrate the whole exercise. There may be a deadline for action. If you are designing equipment specifically for use during a forthcoming eclipse you will not need to be reminded that it will be better to have an adequate device set up in time than the 'optimum' – if that could indeed ever be recognised – ten seconds late. Timing often presents an acute problem in this context. You are researching in a competitive field, you have enough to publish but suspect that a really significant discovery is just round the corner. If you publish now you may alert the competition, but if you delay and they publish first you lose almost all. How long to wait? This problem is a very close analogue of many that are encountered in industry.

In the last resort many important decisions have to be made as a matter of judgment. The outcome cannot be predicted with certainty on the basis of available information or analysis and, when there is no time or no case for further investigation, the decision maker must try to shorten the odds against him by drawing on experience, wit, wisdom and the other ingredients of flair or hunch. But only if decisions are taken with due regard for all the information that can reasonably be made available, can he improve best guess into informed judgment.

In summary, there are three questions to be answered at this stage: Have I a problem and if so what is it? What if anything can I do about it? If there is something that I can do, will it be worthwhile doing anyway?

2.3 **Delineate the system to be studied**

2.3.1 *The model as an abstraction*

A model is by definition an abstraction. The transformation from prototype to model can never result in identity. At best we can make an almost identical copy but usually we need to construct a model that will be significantly different from the prototype; were this not so we would work with the prototype itself. Modelling is a process of selection and transformation: selection, firstly, of what to represent, that is to say defining the prototype, and then of a set of characteristics of the prototype to be incorporated, after appropriate transformation, in the model. There are thus three stages of abstraction at each of which relevance may be lost. Much of this chapter is concerned with the ways of avoiding this loss. The current section deals with the defining of the prototype and with the problems that can arise in seeking to isolate something from its environment in order to study it. Section 2.4 develops this theme further and also takes up the question that is a principle concern in section 2.5; how to identify among the countless characteristics of any system those that are relevant to the purpose in hand. The third abstraction, that inherent in the transformation of these characteristics for representation in the model, is also discussed in section 2.5.

Nothing can be so much as taken out of its environment and remain unchanged. In spite of this, the scientist usually considers it essential to isolate his experiments from the environment in order to work under 'controlled conditions'. He tries to identify all the significant variables that may come into play and to see that they are adequately controlled, or at least that their variations are measured, so that their effects may be identified and quantified. Unrecognised or uncontrolled variables confuse the pattern of results by generating noise which at worst may make them meaningless, and at best will increase the number of experiments that must be done before significant conclusions can be drawn.

In trying to create the clean, well-controlled experimental situation, the scientist runs the risk of unwittingly suppressing relevant factors by drawing the bounds of the area of study too narrowly, or by neglecting important factors within it, or by tackling a complex situation piecemeal in a fragmentary, uncoordinated way. What has become known as 'the systems movement' developed in response to the need to avoid these pitfalls.

2.3.2 *Systems analysis*

Systems analysis (refs. 3 and 4) which has had something of a vogue over recent years and many important successes withal, has as objective to ensure that the totality of that which has a bearing on the problem in hand is included in the study. It may seem inconceivable that anyone should set out to study less than what is necessary and sufficient for the purpose in view. But there was no shortage of error for systems analysis to feed on as it developed in response to the need for a better process for problem solving than the traditional intuitive approach. By way of exemplification of the systems method, we will leave modelling for a moment and turn to a simple every-day problem to illustrate this way of thinking.

A small factory buys in steel parts and assembles them into filing cabinets which are then painted and sold. The paint is sprayed on from compressed air guns. When finished the cabinets are good value for money and the demand for them is rising, so much so that the four spray guns are kept running flat out – or they would be were it not for the fact that from time to time there is a partial failure of the compressed air supply. When this happens the air pressure in the paint gun drops and the paint stream is not properly atomised. In consequence large particles are deposited on the cabinets causing unsightly blobs and runs and producing an unsaleable cabinet. The factory owner recognises that if he were able to maintain air pressure these defects would not arise and so concludes that the best thing to do is to buy a bigger compressor. He decides, however, before finally placing the order, to call in a young engineer who is an expert on compressors as a consultant to make sure that he buys the most suitable type.

Problem as put to engineer – I need a supply of clean compressed air so many cubic feet per hour at so many p.s.i. gauge, what machine do you recommend? System – a compressed air machine operating in isolation.

Requests of this sort frequently come the way of the engineer and scientist. Instead of describing his actual problem, the client reformulates it into a supposedly technical specification of 'the' solution. By offering an inadequate model instead of the real problem he seriously prejudices likelihood that the consultant will find the best solution and indeed the 'problem' as perceived by the client may not represent the actual trouble.

The engineer is a competent professional and has seen all this before. He knows that the best type of pump cannot be chosen on the basis of delivery rate and pressure alone. He knows, for example, that flexibility of output and steadiness of pressure are essential in some applications. So he takes the crucial step and enlarges the system by asking 'Why?' – 'Why do you want to change your compressor?'

'Well, I sell filing cabinets – and I could sell more – but my spray guns keep going haywire because the air pressure drops and when that happens, etc . . . '

The problem has been reformulated and the system enlarged, now it is a pump in a working environment.

'Perhaps I could save you money', says the engineer. 'I bet we can get a better performance out of your existing machine. When did you last have it overhauled? – and look at the dent in this pipe, quite a constriction' – he unscrews a connection – 'Half blocked with grease, too.'

Well the engineer like many scientists is prepared to discuss and hypothesise only for just as long as he can't try something out and, given the chance to experiment, he prefers to resort to his enactive models and to tackle the 'How' (how do we put it right) rather than the 'Why' (why do we need to do something). The next day he takes the compressor to pieces, cleans it and sets to work putting it together again. But he keeps on thinking and when the factory owner drops in to see how he is getting on he envisages a still larger system and makes another suggestion:

'What about running just three spray guns but on shift – wouldn't the increased output with the saving of the cost of a new compressor more than pay for the overtime rates?' The owner agrees that it probably would – here indeed is one solution based on a wider system than that originally envisaged which is possibly superior (provided that overtime working is acceptable to the staff) to what the proprietor first had in mind. But another technical point has occurred to our engineer and a most important one.

'By the way, you said the pressure drops from time to time, I well see from the state this machine was in why it might be low but not why it should fluctuate – any idea?'

'Well, it's usually when we switch on those sand-blasters.'

'You didn't say anything about sand-blasters.'

'No. You didn't ask me. We have two sand-blasters: they're in the next bay to the paint booths and of course they take compressed air too. The men who assemble the cabinets clean the parts up with the sand-blasters, ready for painting, and sometimes the painters have to help out with this job too.'

'I don't see why you should have to put up with dirty pressings. Can't you insist that your suppliers deliver them clean?'

'Oh they are usually all right when we get them, but there is no room to store them under cover so they're stacked in the yard and when it rains they get a bit rusty. Then there are those blobs and runs from the spray gun – I told you – when the air pressure falls – they have to be sand-blasted off and then we repaint.'

The story just summarised is, in fact, the basis of a case study that is used as an operational research exercise. It is taken from a paper by Raybould (ref. 5). Students spend a whole morning on it – taking the role of the engineer, with an instructor playing the part of the factory owner who volunteers information only when asked. The actual problem has convolutions and complications that are irrelevant for present purposes and there are many more solutions than any we have suggested so far, but the reader may like to work out, on the basis of the information before him, what he would recommend.

The final simplification is to scrap the sand-blasters and use the space freed for storing clean parts as received – stored steel pressings, no rusting – no rusting, no need for sand-blasting – no sand-blasting, no fall in air pressure – no fall in air pressure, no blobs and runs – no blobs and runs, no need for sand-blasters.

This is a simple example of the gradual approximation to the totality of factors relevant to a given problem – in other words a move to the relevant system. Even so, many potentially important aspects of the situation have not been mentioned. For example, would the solution with its resultant change in work pattern be acceptable to the employees? Indeed, it would not be too difficult to write a credible scenario that would greatly extend the implications of a decision which seemingly relates only to the small and apparently independent factory. The question has more than once been asked, can we ever set limits to the results of our interventions in a situation?

2.3.3 *The quasi-isolate*

Can systems really be isolated? Theoretically of course they cannot. As Eddington said, if we but move our little finger the distant galaxies tremble. But astronomers do not rewrite the nautical almanac when they learn the date of the Olympic Games. But at shorter range, and especially when people may be affected, action waves may decay slowly and ripples or their reflections may have consequences never dreamt of when an experimental stone was thrown into the pond.

The point is developed by Kelly (ref. 6) who writes

> The universe that we presume to exist has another
> important characteristic: it is integral. By that we mean it
> functions as a single unit with all its imaginable parts
> having an exact relationship to each other. This may, at first
> sight, seem a little implausible, since ordinarily it would
> appear that there is a closer relationship between the
> motion of my fingers and the action of the typewriter keys
> than there is, say, between either of them and the price of
> yak milk in Tibet. But we believe that, in the long run, all
> of these events – the motion of my fingers, the action of the
> keys and the price of yak milk are interlocked. It is only
> within a limited section of the universe, that part we call
> the earth and that span of time we recognise as the present
> eon, that two of these necessarily seem more closely
> related to each other than either of them is to the third.

The extent to which parts of the universe interact can be thought of in terms of a statistical function, the coefficient of correlation. Kelly's argument implies that whatever we do would be seen to affect everything else in the universe in some degree, if only a sufficiently long perspective were taken. For most pairs of events, however, the coefficient of correlation is so small that it can, to all intents and purposes, be neglected. When a system stands in this relationship to the rest of the universe we will describe it as a quasi-isolate, or as being quasi-independent. The qualifier 'quasi' is used because most of the systems with which we have to do are sufficiently close to others that are of importance to us, in space and in time, to make it improbable that they are independent to *all* intents and purposes, but only for *most* intents and purposes.

The scientist believing himself able to isolate all his experiments from their environment for the purposes of research, falls ready victim to the snare of extrapolating this assumption to situations in which it no longer holds true. His training in classical thermodynamics leaves the chemist especially vulnerable, encouraged as he is to think in terms of the attainment of time-independent equilibrium states with maximum entropy and minimum free energy, as required by the second law of thermodynamics, a law which, though widely regarded as one of the most 'fundamental' presupposes the existence of isolated, closed systems. As is well known, the classical laws of thermodynamics proved inapplicable to the behaviour of living organisms, which appear to bring about a reduction in entropy. A number of workers including Denbigh (ref. 7) and Prigogine (ref. 8) have developed thermodynamics of open systems, though even this has not been sufficient to establish an accepted basis for thermodynamics in biological systems, as the controversy over the supposed 'high energy bond' in adenosine triphosphate (ATP) and its role in metabolism bears witness (ref. 9).

Writers still try to make distinctions between open and closed systems as though they represented a real dichotomy. It would be better to adopt Kelly's standpoint and accept that *no* systems are closed and that the question to be asked in any situation is whether the system may be regarded as *closed for the purposes in mind*. In other words whether, for those purposes, we may consider it a quasi-isolate. It is, therefore, for each experimenter to decide whether he treats his system as closed and, if he does, to remember that he is dealing at best with a quasi-isolate and to think out what limitations this places on the scope and relevance of his results.

The existence of systems that are quasi-isolates is not merely an experimental convenience but is a matter of practical necessity. The homeostasis of living things is an expression of the need of organisms, in the pursuit of self-preservation, to insulate themselves from the effects of many environmental changes. Biological evolution would not be possible if all species interacted so richly that all had to change at the same time. There must be some decoupling of units within a system that is to evolve. A chemical plant in which there is no interstage storage, in which one reactor feeds directly into another can only be operated over a very narrow range of conditions. Either everything works as planned or the whole system seizes up. Uncoupling devices (e.g. intermediate storage) to restore independence of operation to the various stages are expensive but it may be vastly more expensive to do without them.

Quasi-isolates are necessary in large social organisations. For example, in industry the top brass of large firms often seek to improve performance by 'optimising' some component over all businesses in their group. An optimal group research strategy, or computer, or distribution, or maintenance strategy will be sought and imposed across all units. Now, of course, there is synergy to be found in many of these operations and there are better and worse ways of conducting them in a corporate sense. But synergy can be bought only at the cost of degrees of freedom, freedom that is necessary not simply for the expression of individuality but for the adaptation and evolution that is essential to continued viability in a changing environment.

To return to biological evolution, changes in one species do, of course, demand adaptation in others and these will occur provided that the rate of change is not too great and that there is time to adapt. But there is not always enough time. A mutation in an influenza virus to give a lethal strain may wipe out populations. On the inorganic level, earthquakes and volcanic eruptions are events that create pressure on their immediate environments too rapidly to permit adaptation. Such cataclysmic events are now well known also on a cosmic scale.

So much for the stage of defining the system, but all activities in modelling interact and interpenetrate and the systems concept will remain very much in evidence in subsequent sections. So far, we have concentrated attention on the first of three stages of abstraction that are inevitable in modelling; to wit, the selection of a set of events, objects, interactions, processes in a limited span of time and space as prototype for our model or as defining the task environment in the real world. In doing so we inevitably postulate a quasi-independence but if the system is drawn too narrowly there will be at best dissatisfaction and at worst disaster.

The converse error of drawing the system too widely can lead to waste of resource in the study and in the solution, as well as to the appropriation of degrees of freedom that should be left for others.

2.4 Formulate questions to be asked of the model

2.4.1 *Types of experimentation*

The aphorism 'Ask the right questions and the answers will take care of themselves' is true for most experimentation. The 'right' questions, in our context, will provide answers that reduce the uncertainty that contributes to our having a problem; the questions will also help to direct attention to those properties of the prototype that must be represented in the model.

Experimentation has, broadly speaking, either general or specific objectives and each requires a particular approach to question formulation. The first, the general objective, is just to take a look; out of sheer curiosity in the hope that something interesting or important may be spotted or as a reconnaissance – getting the feet wet. It may not be possible at this stage to formulate precisely what questions the experiments might answer. The essence of this approach is to get to grips with a situation using whatever resources lie close at hand and accepting such limitations as cannot be easily and quickly removed. This may be done by constructing a simple model from readily available information and with free recourse, when necessary, to approximation and informed guess-work. It is usually good practice to begin in this way. Start simply and with a broad sweep and add detail later is a good precept.

The second type of traditional experimentation is one in which specific questions may be formulated in advance, though there may, of course, be many surprises in the answers. The researcher is seeking to elucidate or to add to the structure of scientific knowledge, to establish connections or to quantify. Such activity may range from simply filling in the picture to the development of new hypothesis and theory. There is a continuous spectrum of significance though it is never possible to be sure at the outset whether one's results will lie closer to the trivial or to the highly significant. The odds will be biased by the imagination of the researcher. All experimentation seeks to increase understanding of a system. Increasing understanding implies the recognition of a richer pattern of relationships in the internal environment of the system under study or between it and its external environment.

The applied scientist may well be far less free to choose his problem than his academic colleague. In pure research every discovery is likely to be publishable whether or not it relates to the problem as envisaged at the outset. But if you have to design a bridge, then a novel motor car, however good, will not be an acceptable outcome. Moreover, the implications of a misfit with the environment are usually very different in academic work from those which arise in the applied sciences, which are largely the province of modelling in industry. In pure science the tests that must be used to determine whether one's results are compatible with the environment – the existing corpus of knowledge – are clear and well understood. Moreover, in the rare but important instances in which experimental results require the revision of accepted ideas, the necessary adjustments can often be fairly readily identified. On the other hand, the world into which must fit the designs and intentions that are developed as a result of modelling in applied science is poorly understood, ill-structured, stochastic, very variable with time and susceptible to unpredictable human intervention.

In laboratory studies the unexpected outcome is often welcome as opening up new lines of thought. The same welcome may be extended to the unexpected event that is encountered when modelling for a practical prototype, provided that the surprise occurs with the model and not first with the prototype. When experimentation has as its objective the expanding of scientific knowledge then it is, as we said, into this corpus of knowledge that the new findings must fit. Even if they are not really at home and are ultimately rejected, no harm, other than to reputations, is likely to be done. But in practical, everyday situations rejection of a misfit may be catastrophic.

This need to meet an increasingly demanding specification including social and environmental considerations that might in former times have been ignored is potentially either a challenge or a source of inhibition. Must it mean, as Wylie Sypher suggests (ref. 10) that the technologist, in his dread of being surprised, predicts everything and discovers nothing? This is a question to which we return at greater length later. Here we merely note that the technologist's task is not to implement, by some mechanical procedure, a plan that is fully illuminated by another's flash of scientific inspiration. It is to express and realise possibilities, often obscure, implicit in germinal observations and this task, if well done, calls for a large measure of creative imagination.

2.4.2 *Targetted experimentation*

Let us discuss first the second type of experimentation, which has the more specific objectives. Modelling for design purposes is clearly of this type. Suppose a satellite is being designed. It is likely that several distinct iconic models will be made; to investigate mechanical strength, heat resistance, telemetry, meteorite impact, postural stability and so on. Consider the model that is to be built to assist in ensuring that the prototype has adequate mechanical strength. What is adequate will depend on the degree of deformation that can be tolerated, the lifetime that is expected and the nature of the task environment. Already we see that important questions are being generated; what deflection *can* be tolerated? What is the working life to be? And in considering the task environment it must be remembered that the satellite will have to withstand the stresses to which it is subjected not only during its orbits in space, but during the launch in which the stresses imposed by the environment, for example temperature and acceleration, will be quite different.

When all these matters have been thought out, it will be possible to define what uncertainties exist and so to formulate questions to be answered by the experiment. They will be of the form 'What changes (e.g. damage, deformation) occur in the model after exposure to specified stresses, e.g. vibration, temperature, impact for a given time'. Since the extent of change that the satellite could suffer and remain operable will be estimable, the answers to such questions will relate directly to the suitability of the design that is under test.

A precise definition of the experimental task in terms of questions such as these has numerous advantages over the more traditional and vaguer statement, of the kind that 'the objective of the work is to study factors affecting the mechanical strength of a satellite' among the more important of which are the following.

(1) Attention is drawn at an early stage to the importance of defining the task environments (physical, social, economic or whatever) over the whole life of the prototype, of considering what parameters will define satisfactory performance. A clear statement as to what questions the modelling is to answer constitutes an important first step in ensuring that the results of the model experiments can be translated meaningfully into the real life situation.

(2) It specifies the task in terms that are convenient for critical questioning, for example, by the process of critical examination which consists of questioning the objective systematically in a way that generates alternatives.

(3) It helps to avoid one of the biggest time wasters in experimentation – the discovery, late in the day, that the pattern of experiments is bad from the point of view of the statistical analysis that will quantify the significance of the results and the confidence that can be placed in them.

2.4.3 *Exploratory experimentation*

So much for experimentation of the second type, but how does one formulate questions relevant to experimentation of type one, which is less focussed and often in the nature of a reconnaissance into quite unknown territory?

In chapter 6 in discussing the modelling of enzyme systems to study their mode of action, it is suggested that the first step is usually a general exploration of the behaviour of the enzyme by use of such techniques as are readily available. For example, is the enzyme inhibited by alkylating agents or is the presence of a particular metal ion important? Or, how is the rate of the enzyme catalysed reaction influenced by pH?

The answers to these questions, which relate here to pre-modelling experimentation, will help to locate the enzyme within the experimenter's existing conceptual model of enzymes, and thus suggest the way ahead – that some things are possible and others not.

Or to turn to a less obvious example, an industrial toxicological laboratory has, as one of its tasks, the testing of chemicals for skin irritancy; so that appropriate precautions can be defined for any that are potentially harmful. The usual test is to apply the chemical to shaved areas of rats' skin and to keep watch, over a period of time, for adverse effects. The test is reasonably satisfactory in that rat skin seems to be a good model of human skin, in so far as compounds that are harmful to man appear to be harmful to rats also. But the test uses and hurts a lot of animals and gives results that are not easily quantified and moreover, the nature of the lesions produced are not well understood. The problem seems to be that not enough is known about the mechanisms by which the skin can be damaged. It might be a good idea to study the tissue changes that occur using isolated dermal cells, that is to say, the living layer beneath the epidermis and, indeed, there is reason to believe that the measurement of the changes in concentration of certain enzymes in dermal tissue when chemicals are applied to the outer skin could provide a quantitative indication of the extent of damage. If this proved possible we might have an improved model by which to test chemicals. And so the proposal might be 'to study the effect of a range of chemicals, known to be harmful to human skin, on the specified enzyme systems in isolated dermal cells from the rat '. But this is a vague proposal that will suffice only if information of itself, any information, is considered valuable. Can an objective be defined more closely?

If that objective is to devise a test that will be less disagreeable, more accurate and more economical than that currently used with live rats, then questions can be formulated that relate to correspondence between the changes in the enzyme systems that can be observed in the new test with the skin changes or lesions that are produced in the test with live rats. And in this case we will rely on our existing knowledge of the relevance of the tests with rats to provide a link with real world situations. If, however, what we are looking for is something that will provide better guidance as to safe behaviour, then the objective is different and we must ask in what situations the information derived from our tests, that is to say from our model experiment, is likely to be put to use.

There are essentially two types of potentially dangerous exposure; a single exposure to a sufficient concentration of chemical to produce damage – a situation described as acute toxicity – and also the possibility that long-term exposures to much lower concentrations may give rise to disorders, in other words, that there may be some chronic toxicity. The outer skin is, of course, fairly resistant to many chemicals and most of the compounds that we will want to test will have to penetrate the skin before they can cause any damage. The ease with which the skin can be penetrated will obviously depend on the presence or absence of cuts. Moreover, it has been observed in some experiments that the state of the hair follicles, which may go through a cyclic pattern, is an important factor in the resistance of skin penetration by applied chemicals.

We will need to consider in what practical situations the chemicals we are testing are likely to be used. Will protective clothing be worn thus possibly setting up damp areas which might affect the translocation of the chemicals? Is the chemical likely to come into contact with skin in solution and, if so, in what, or as a solid or as a dust or how?

By thinking through the practical situations in which the chemicals under test will be used, we can decide what toxicological information would be useful. Questions which, if answered, would provide this information can be formulated and we can then see whether the proposed experiments with isolated cells are likely to answer them; if not, they may be irrelevant. The important point is that a decision on these matters will have to be taken before the results of the proposed work can be of practical use. Focussing attention early on what needs to be known greatly increases the probability of getting meaningful results.

2.5 Constructing the model

In this book we draw attention to some principles that underlie all these activities and try to show their relevance in practical situations. However, it is not the objective to give detailed advice on the wide variety of skills and procedures that are needed, though several examples of models are treated in some detail and references are given throughout this book to texts that treat *in extenso* particular aspects of modelling.

2.5.1 *Identifying parameters*

Models are made in many shapes and guises; a concept of how things might happen and why; a map; wooden blocks, tubing, cardboard representing anything from a design for a stage set to a design for a chemical plant; a semi-technical plant; an aeroplane mock-up; an isotopically labelled molecule; a symbolic description perhaps in mathematical form, of almost anything.

In each case it is necessary to identify those characteristics of the prototype – its properties, relationships, behaviour – that are to be represented in the model. Often this will seem a formidable task. Systems, especially in real life, may have millions of variables and there is no rigorous method of recognising which of them *are* important for the purposes in hand.

Fortunately, in any given situation, it is rarely the case that we must be concerned with *all* the detail of the phenomenon that we are studying; there are usually only a few properties embedded in the complex of reality that are relevant. Simon (ref. 11) dealing with the simulation of poorly understood systems makes the point very clearly.

> The more we are willing to abstract from the detail of a set of phenomena, the easier it becomes to simulate the phenomena. Moreover, we do not have to know, or guess at, all the internal structure of a system, but only that part of it that is crucial to the abstraction.

And he continues with observations that are relevant to hierarchies and to empirical and theoretical models – both of which will be discussed in subsequent sections of this chapter.

> It is fortunate that this is so, for if it were not, the top down strategy that built the natural sciences over the past three centuries would have been infeasible. We knew a great deal about the gross physical and chemical behaviour of matter before we had a knowledge of molecules, a great deal about molecular chemistry before we had an atomic theory, and a great deal about atoms before we had any theory of elementary particles – if, indeed, we have such a theory today.
>
> This skyhook-skyscraper construction of science from the roof down to the yet unconstructed foundations was possible because the behaviour of the system at each level depended on only a very approximate, simplified, abstracted characterisation of the system at the level next beneath. This is lucky, else the safety of bridges and airplanes might depend on the '8-fold way' of looking at elementary particles.

Even though it may be necessary to represent only a relatively small proportion of the properties of a system, the ramifications are often so great that it is well nigh impossible to make, initially at least, a single model of the whole. The reason lies partly in the system and partly in ourselves. The complexity that the mind can encompass, even with the aid of the convenient symbolism that modelling should offer, is limited. A set of interlocking sub-systems will be far easier to grasp than the unsegmented totality. Moreover, if the total, complex system is modelled all of a piece, then the number of interactions that have to be represented may become very large indeed. The result of this proliferation may be not merely complexity but rigidity, if each part of the system has to adjust to a change in each other part.

What is needed is a set of quasi-independent sub-systems each of which can be modelled in the most appropriate way and optimised independently, and each of which is so structured as to fit together, eventually into a hierarchy of systems, the output from one level of which is an input of the next.

A useful way of seeking out significant variables and a related method for decomposing systems into quasi-independent sub-systems will be found in a most interesting paperback written by the architect Christopher Alexander, *Notes on the Synthesis of Form* (ref. 12) which is strongly recommended reading. Alexander's principal theme is that a diagram (an iconic model) can be a very powerful tool for teasing out the complex web of a design problem and for suggesting practical solutions, and in developing this theme he has much to say that is very relevant to modelling. The following paragraphs summarise some of his arguments and relate them to parallel concepts that are presented in this book.

Alexander begins by contrasting the process of design in primitive societies, where it is unselfconscious and learnt by following example, with the conscious struggle of the modern designer who has great difficulty in creating forms that are not at odds with their environment (see also ref. 13). The unselfconscious designer, on the other hand, producing as he does artifacts that copy others that have traditionally proved their suitability for the purposes in mind, generally succeeds in avoiding misfits between his creations and their environment. That this should be the case is perhaps not at all surprising when one considers the potential power of modern artifacts, the rapidly changing situation in which they have to function, and the great complexity and range of interaction of modern society.

Be that as it may, Alexander draws from his antithesis the concept of 'misfit with the environment'. The avoidance of misfit between the object designed which he calls the form, and its task environment which he calls the context is, according to Alexander, the most important task in design. The form is the part of the world over which we have control. The context is any part of the world that puts demands on this form. Fitness is a relationship of mutual acceptability between form and context.

The search for misfits is a key concept in Alexander's thesis. Indeed he maintains that it is virtually impossible to define objectives by reference to suitability rather than to incompatibility. Though we would argue that this is not strictly true, there is no doubt that the concept of misfit is both practical and very powerful. Take for example a relatively simple task such as painting a wall. We *could* define a satisfactory finished job in terms of the reflectance of light of specified wavelengths from all points on the surface. But if we were concerned with good covering of the existing surface, we would be more likely to think in terms of the unsatisfactory outcome, i.e. the misfit that reveals itself when the underlayer shows, or 'grins' through as they say in the trade. It is essential to note that a misfit can be identified only in relation to both form and context together. Insofar as people construe contexts and expectations differently, there may be disagreement as to what is a misfit and what is acceptable. If, in respect of a particular feature of a design, it is not possible to imagine a misfit then it need not be considered further.

The possibility of mismatch with the task environment and its early detection is, of course, at the heart of the screening process to which many industrial products are subjected. We shall see in chapter 8 that the screens used in evaluating new products are essentially models of the task environment in which candidate products will have to perform. Screening is essential when, as is usually the case, testing in the real world situation is not possible. Sometimes there are measurable characteristics of the products that can be related to their performance in the task environment. In such cases the test procedure for measuring these characteristics can serve, in this respect, as a model of the environment and a predictor of eventual goodness of fit.

The first task in the design process is, according to Alexander, to draw up as full a list as possible of potential misfits which serves as 'a temporary catalogue of those errors that seem to need correction'. The definition of these potential misfits is important, as they will provide the basis for the decomposition of the design problem into a hierarchical system of sub-problems in a way that will structure the design programme. This procedure is closely akin to that of defining questions to be asked of the model. For example, set the general but ill-defined task of designing a distillation column suitable for the separation of the components of a particular mixture, the designers consider specifically what might go wrong; corrosion of the column, decomposition or polymerisation of various components, channelling of reflux through the packing, collapse of the packing because of its own weight on the lower layers and so on.

Having listed all the misfits that can be thought of, one then proceeds to look for interactions between them. Misfits interfere or interact when action necessary to deal with the one makes it more difficult – or easier – to deal with the other. For example, in designing an electic light bulb, amongst the conceivable misfits are that the lamp burns out too soon and that the lamp does not emit enough light. One way of increasing brightness would be to run the filament hotter, but this is likely to reduce filament life, and so these two misfits interact.

As anyone who has tried to develop a new product knows, a major problem is how to achieve an acceptable compromise between conflicting demands – how to resolve the interactions between misfits of the product with its task environment. Alexander offers help in this direction by proposing a way of structuring the interacting misfits into sets that will assist the designer to see how a complex design problem can be tackled piece by piece; how, in fact, a complex system may be broken down into a hierarchy of what we would call quasi-independent sub-systems. The argument is too complex to be summarised here but well worth reading in the original.

The germinal concept to which all Alexander's analysis leads (such is the way his book is written) is that of expressing the requirements of a problem and the possible solutions to it in diagrams, which are defined as 'any pattern which, by being abstracted from a real situation, conveys the physical influences of certain demands or forces' (ref. 12a). If the patterns are a notation for the problem itself, then they are requirement diagrams, if they represent the forms which might be created to respond to the demands expressed by the requirement diagram then they are form diagrams.

Requirement diagrams are useful only when they contain physical implications and so a measure of form. Diagrams which relate both to requirement and to form are called, by Alexander, constructive diagrams. What a thing is, in itself and in isolation is its formal description; what it does when introduced into a real world environment is its functional description. When we do not understand objects well enough it is very difficult to relate the two descriptions. When we thoroughly understand the situation then the two descriptions are equivalent and we have a unified description which is the abstract equivalent of the constructive diagram.

> The texture of bathers on a crowded bathing beach is a diagram. The evenness of the texture tells you there are forces tending to place family groups as far as possible (and hence at equal distances) from one another, instead of allowing them to place themselves randomly . . . Kekulé's representation of the benzene molecule (as atoms with linear bonds between them) is again a diagram. Given the valency forces represented by the bonds, the diagram expresses the physical arrangement of the atoms, relative to one another, which is thought to result from the interaction of these valencies. [Ref. 12a.]

As we know, and as we shall discuss in greater detail in chapter 5, Kekulé's representation expresses a good deal more, it is in fact, to some extent a functional as well as a formal description. For example, the double bonds indicate a degree of activity greater than would be found in paraffins, but the formula is misleading in implying that an olefinic type of reaction is possible. Alexander considers that one of the tasks of the designer is to bring the formal and functional descriptions into coincidence and we will see later this process in progress in relation to the benzene molecule.

Why has chemistry, as Alexander avers, been particularly successful in bringing its formal and functional descriptions into coincidence? Perhaps because the sub-systems (e.g. molecules) are quasi-independent, being physically independent to a large degree because considerable energy is normally required to bring them into interaction. Form, in chemistry, is very directly related to function as the concept of 'functional groups', which will be discussed in chapter 3, underlines. Moreover, it is usually relatively easy to define the environment in chemical systems, so that one source of uncertainty as to the relationship between form and function is controlled. Of course, in complex systems such as are met in biochemistry, the environment of any given reactive centre may be far from well defined and we shall see the implications of this when discussing modelling in biochemistry in chapter 6.

Because of the relative ease with which form can be related to function, one can go a long way in chemistry by intuitive, even unselfconscious modelling and before the development of modern techniques for structural analysis, such as nuclear magnetic resonance and electronic spin resonance, it was all too easy not to think beyond an empirical level. The new techniques have forced a breach in the walls round the old thinking because their proper language is more theoretical and therefore more general.

2.5.2 *Hierarchies*

Problems, whether of design or otherwise, are often far too complex to be solved all in one piece and need to be broken down into components which, though interlinked, are quasi-independent and can be treated as such for the purpose of modelling. In this way a multi-component problem, very difficult to hold in the mind, is converted into a number of smaller, manageable problems which may themselves be further sub-divided to produce a pyramid-like hierarchy.

In this type of hierarchy each higher level is more comprehensive than the next lower. The higher model includes a part of the external environment of the lower and, therefore, if it is well constructed will provide information relevant to a correspondingly larger slice of the universe. The higher may, however, have no more *generality* than the lower model, it may be no more *fundamental*. But a hierarchy of models may also be constructed in which each higher level is expressed in more fundamental terms and has, therefore, a greater generality. The chemist's conceptual models that are discussed at length in chapters 3 and 5 form such a hierarchy and a brief summary of part of the argument that will be developed later will indicate what is meant.

In principle, all the phenomena of chemistry could be deduced from the properties of electrons, protons, neutrons and photons, since the first three mentioned are the fundamental particles of which all matter is made and for practical terrestrial purposes, we may regard photons as the only particulate form of energy. Quantum mechanics provides a mathematical model that can express all the necessary interactions but, unfortunately, the equations are so complex that they can be solved only for the simplest molecules. How this difficulty is partially circumvented will be seen in chapter 3.

Since the fundamental quantum mechanical approach is restricted in its application, the chemist has recourse to more specific models, in particular those that are constructed in terms of the stable groupings of fundamental particles, that is to say, for the inorganic chemist principally elements, and for the organic chemist groups of atoms that exhibit particular patterns of properties and which are known as functional groups. Models at the functional group level are more precise but less general than those at the quantum mechanical level.

The class of models based on functional groups began by assuming tacitly that a given group, say amino $-NH_2$ or carboxy $-COOH$, would have the same properties wherever it occurred in a molecule. In other words, the molecular environment in which the functional group found itself was ignored. However, although the conceptual model that represented the chemical properties of a molecule as essentially the sum of those of its constituent functional groups was remarkably successful, it became apparent that the molecular environment of a group could markedly affect its reactivity, often qualitatively and sometimes quantitatively. And so a third class of model, still more specific, was introduced in which 'effects' were postulated to explain such environmental interactions. There is thus yet another level of still greater particularity in the chemist's kit of models.

In this hierarchy each higher level 'explains' the lower, in that the same phenomena are expressed in terms of concepts that have a greater generality, and thus integrated more fully into the corpus of scientific knowledge. From the higher standpoint in the hierarchy, the lower level models are more 'empirical' and conversely, the higher are the more 'theoretical'. The significance for modelling of these two concepts will be examined in the next section.

2.5.3 *Empirical and theoretical models*

One way of setting up a mathematical model of a particular system, say a reactor in a chemical process, is to study its behaviour and derive a model that correlates process variables that can be controlled by the operator (e.g. reactor temperature, reactant feed rate etc.) with output variables such as composition of product. In one such procedure, linear regression analysis (ref. 14), it is necessary to identify by experiment, experience or intuition those factors in the system that affect its behaviour significantly and then to run the process under a series of conditions that cover a range of values for each factor, noting the values of the output variables in each case. Regression analysis produces a set of equations that relate the input and output variables statistically. It is also possible to calculate how much of the variance in the process behaviour can be significantly attributed to the specified process variables, so that the existence of any unsuspected significant variable should be exposed.

Such a set of equations is known as an empirical model because it is derived directly from experiment and owes nothing – except possibly the identification of some significant variables – to theory. The equations may take any convenient form and not necessarily those forms commonly used in chemistry and chemical engineering. Empirical models are often a very good representation of a system since they encompass not only the underlying laws that are of general application but also the idiosyncracies of the particular prototype, and can be expressed in terms of readily measured parameters.

On the other hand, the empirical model is inevitably limited in its applicability because it is a special case. Like a topographical map of a stretch of country it is purely descriptive and does not concern itself with causes. Though it may represent very precisely the accidents of the terrain within the area depicted it will usually be of little use either in explaining underlying structure or in predicting what happens outside its boundaries. Discontinuities may be improbable – a river valley may be expected to continue on the next sheet – but the possibility of a geological fault cannot be excluded. The further away one moves the greater the uncertainty. Even within its boundaries there may be discontinuities; surface features of height less than the contour interval may not be indicated, but a forty-nine-foot drop can be very disconcerting. So it is with empirical models, extrapolation is very risky and even interpolation demands care.

An alternative course is to construct a mathematical model in which the equations are in the form of accepted laws and relationships. For example, the chemical engineer who is designing a reactor might build a theoretical model in which the Arrhenius equation is used to express the dependence of reaction rate on temperature, and so on. Such a model has the advantage that it relates the particular case under study to scientific theory in general. It enables the scientist to model in a precise and widely understood language. The equations which are used in such a theoretical model are themselves well tried and tested models – they are written in a code that is well understood by scientists even if they initially did not know what all the terms in the code meant (e.g. *A* and *E* factors in the Arrhenius equation). Of course, not knowing what parameters mean, sounds very much like working with an empirical model and indeed all mathematical models, be they designated theoretical or not, ultimately derive from experiment, i.e. from empirical data. The distinction between theoretical and empirical, in this context, is that the former is written in a language that is applicable to systems larger than the one currently under consideration. Thus, if you have a theoretical model of an operating plant and something goes wrong then you are ready poised to call in aid the whole corpus of scientific theory, the bridges are already built. An empirical model on the other hand, since it is a description of behaviour in terms of the special case, provides no easy way of relating the unexpected aberration to the underlying cause.

Some processes, for example those in which turbulence is important as in the atmospheric systems that control the weather and some hydraulic and aerodynamic systems, are too complex to be modelled theoretically and an empirical approach is the only possibility. Likewise, if the reaction referred to in the preceding paragraphs were catalysed heterogenously, the complexity of reactant and product diffusion into and out of pores, thermal gradients, surface inhomogeneities, poisoning and so on, would probably dictate an empirical approach.

As we shall see in chapter 7, in developing chemical processes it may be best to design the plant and, perhaps, design and install computer control on the basis of an essentially theoretical model and then, as experience at full scale grows, build a more compact empirical model which should be more economical in computer usage and probably represent the behaviour of the plant more accurately.

2.5.4 *The form of the model*

By this stage of developing the model it will no doubt be clear whether a material (iconic) or conceptual model will be needed, even if this was not apparent at the outset.

The advantages of iconic models are that, because they 'look like' their prototype in some respects there is a visual analogy that assists comprehension. Moreover, even when enough is known to permit the construction of a mathematical model, the iconic model may be richer in preserving properties and relationships of which the importance is not readily recognised. On the other hand, it is less readily apparent with an iconic model precisely what relationships are being preserved, and as we shall see in chapter 7 it may be difficult to find the right scale factors.

Mathematical models make explicit algebraic statements about properties, relationships and limiting conditions. The limitations may, therefore, be more obvious in a particular model and so less potentially misleading than may be the case with the corresponding physical models. The expense is in acquiring the necessary understanding of the system, the problem of setting up the algebraic statements, i.e. of building the mathematical model is usually trivial. Mathematical models can be very flexible, changes can be made simply and quickly. Their operational time scale is usually very short, that is to say they run fast. Storage is no problem and provided that an adequate users description has been prepared they can be readily dusted down and re-used.

The form that an iconic model will take will be, to a large extent, dictated by the fact that it is to be, in some way, a physical representation of the prototype, but if we are to make a symbolic model –

especially if it is to be mathematical – then the algebraic form that it should take may not be obvious. This is a problem on two levels, the simpler relating to how to set up the necessary algebraic relationships when a conceptual description of the model has been achieved and the other, more difficult, relating to arriving at a first conceptual description. The minor problem will readily be coped with by a competent mathematician. There are many off-the-peg procedures available for a variety of problem structures, such as optimisation of a variable under constraints, and for dealing with random (stochastic) as well as with deterministic situations. The second, major problem on the other hand, goes to the heart of many a modelling exercise in that it relates to finding a conceptual description for a seemingly unstructured situation.

In such a case a frequently helpful procedure is to look for a model of a scientific situation that appears to have the required form for the problem in hand. The analogy may be very close as, for example, that between the vibrations of atoms in a molecule and the vibrations of a stretched spring, or it may be distant. In the hydrogen molecule the protons jiggle about but maintain an average distance apart – the bond length – which refers to a molecular configuration of the vibrating system at a potential energy minimum. The potential energy of the molecule is raised whenever the bond is stretched or compressed, very much as in the case of a vibrating spring connecting two weights. The coulombic and exchange forces in the molecule correspond to the restoring force in the spring and the two systems can be formally represented by similar algebraic expressions. (For an oscillating bond, this takes the form:

$$\text{frequency of oscillation} = \frac{1}{2\pi c} \left[\frac{f}{m_a m_b / m_a + m_b} \right]^{\frac{1}{2}}$$

where c is the velocity of light, f is the force constant of the bond, and m_a and m_b are the masses at each end of the bond.)

A more distant example is the visualisation of the change in attitude of a population, away from accepted norms, in response to pressure of information and a subsequent reversal of opinion back again as analogous to a hysteresis loop. There is evidence that if an individual's attitude to a certain proposition that is regarded as a social norm is measured over a period during which he is under pressure – for example by political or commercial publicity to change his mind (let us say for convenience from left to right without ascribing significance to these terms) – then there is likely to be a step change from a slowly increasing dissatisfaction with the current view to a complete reversal of opinion, left to right. The new opinion is then slowly reinforced by continuing pressure. If pressure is reversed then the change back will not occur until his opinion has moved to a point further to the left than he was at when he switched to the right. This suggests that the behaviour of a population under such pressure might follow a course analogous to that which occurs in a bar of iron in a reversing magnetic field – in other words a hysteresis loop.

Models may be static, in which case the relations expressed within them do not change with time, or they may be dynamic in which case this limitation no longer applies. Chemistry is, of course, essentially the study of changes in composition of matter, and kinetics is a fundamental part of a chemist's studies. Even systems intended to remain in steady state are prone to change with time, especially on the industrial scale: catalysts become poisoned, heat transfer surfaces foul up, impurities accumulate. Even when designing a system that is to operate in a fundamentally steady state, one must consider the dynamic problem of how to reach and how to leave it. Every plant at start up must move from ambient to operating conditions and it may well be the case that in doing so it will move into a dangerous state if the modelling and design work are limited to the supposed operating conditions.

Beer sums it up very well (ref. 15).

> The kinds of systems under discussion exhibit literally billions of variables. There is no *rigorous* means of knowing which 'matter'. Indeed the importance of a particular variable in such a system is a question of degree, a question of judgement, a question of convention. Moreover, the importance it has by any of these criteria will change from moment to moment. This does not mean merely that the numerical value assumed by the variable is changing – that is in the nature of variables and one of the things about the system that we know how to handle. No, it means more: *the structural relevance* of the variable inside the system is changing with time.

2.6 **Run the model, analyse the results and their implications**

The running of a model is essentially an experiment and the application of the general principles of good experimental practice will go far to ensuring its success. There is, however, one requirement that is particularly important and that is to make sure that the model *does* represent its prototype for all the purposes in hand. This, the problem of validation or verification has received much attention in the literature; a review and leading references will be found in a paper by Mihram (ref. 16).

The terms 'verification' and 'validation' can usefully be distinguished as Fishman and Kiviat (ref. 17) have suggested. Verification is making sure that the model behaves as the experimenter or model builder intended. If a computer programme is used then debugging, that is to say eliminating errors in the meaning and structure of the programme, will obviously be one component of the process. But even when the semantics and syntax of the programme are faultless, the model will be unsatisfactory if the experimenter does not fully understand what the programme does and, in particular, what limitations are built into it. You do not need to be a mechanic to drive a car but it is useful to know that it is not intended for crossing rivers. The dangers inherent in using a programme as a 'black box', that is to say a device that transforms data in a way that is not clear to the experimenter will be discussed shortly. Verification should, of course, be part and parcel of every stage of modelling.

Validation is the testing of the agreement between the model and the prototype. As such it obviously applies to models that simulate existing prototypes. Since simulations are usually expected to predict, the question is how to gain confidence in the model's predictions. If no harm is done by acting on false predictions, then validation is easy, one can proceed by trial and error. It may be possible to adjust experimental conditions in order to work in this way. For example, in a procedure known as evolutionary operation (EVOP) (ref. 18) for optimising the performance of chemical plant, an empirical model is constructed by establishing a statistical correlation between the parameters that one wishes to improve, e.g. output rate or product quality, and variables in plant operating conditions that are presumed to be relevant. This correlation, a model of the process, is then used to predict by how much specific operating conditions could be varied without making performance unacceptable if the change has an adverse rather than a beneficial effect, and so to move by small, safe steps, doing enough experiments around each new operating point to establish a firm statistical basis for validation before changing conditions further.

Sometimes trial and error validation is not possible because time is too short or because the results of actions taken on the basis of the prediction will pre-empt other alternatives or, if the prediction is wrong, be too damaging. There is not always a comfortable way forward. Important decisions often have to be taken when the outcome is very uncertain. If the uncertainty is too great then we have a problem and re-enter at section 2.2 of this chapter.

To return to the question of safety in practice, one important possibility with the physical model is that of running it to destruction. That is to say that, with appropriate precautions, the model can be overstressed so that its robustness, the extent of its ability to perform in a demanding environment, can be assessed. Such a procedure is, of course, standard practice in setting up safe working limits. It is almost always desirable to extrapolate operating conditions on the model beyond those that correspond to the range of expected working conditions of the prototype. When laboratory results are quickly reproduced on semi-technical plant there is a temptation to cheer and proceed quickly to full scale. But to a blind man, firm ground may feel just the same whether it be a cliff's edge or in the middle of a football field, and a researcher who has only a narrowly based empirical model to guide him will have, at best, very limited vision. It is always prudent, if at all possible, not only to study conditions under which things go well but to explore out until they begin to go badly. The unexpected trouble often lurks close at hand.

It is unnecessary to write at length on the analysis and interpretation of results for which, as for running the model, the principles applicable to experimentation in general are relevant. Results from a modelling exercise should have relevance both for the improvement and development of the model and for the problem that they are intended to illuminate, and they should be so used. It is often possible to use results from early runs to explore the extent to which further development of the model may be helpful. Provided that the limitations of accuracy are borne in mind, much can be deduced from approximate results.

Many scientists seem to believe that imprecise results are useless. Most businessmen believe that all scientists think this way and that because of their aversion to the imprecise, scientists are incapable of exercising 'judgment'. Approximate quantification is, in fact, often not only helpful but also sufficient. Much time is wasted on the unnecessary refining of data. The order of accuracy that is necessary in a particular situation can only be ascertained by examination of the special case in question. Of course, the essence may be not only the possible numerical range in which a parameter may fall but also the probability that it will fall therein. We may be able to tolerate a large uncertainty provided that we can be sure of the limits; indeed the only way to be sure of limits is, often, to accept a wide possible range for the parameter in question. Skill in using imprecise data and taking decisions under uncertainty is a very important attribute of the versatile scientist.

There is indeed no escape from uncertainty, it can best be disarmed by being recognised. Loasby (ref. 19a) warns

> Models whether conceptual or experimental, can be invaluable if their limitations are recognised; but one of the dangers in their use is that they leave us ignorant of our own ignorance. They not only tell us nothing about the effects of what is excluded; they are liable to prevent any recognition that what is excluded may have some effect,

and (ref. 19b)

> One might postulate a law that the sum of rigour is a constant: Thus the greater the rigour of the formal model, the weaker the connection between that model and the reality which it may be used to interpret. The price of precision is not only error, but ignorance.

One might add that the price of rigour and of precision is limited relevance since precise statements can usually be made only about strictly defined systems.

2.7 Doubts and pitfalls
2.7.1 *Technism*

Models and the systems approach are presented in this book with a degree of enthusiasm. That enthusiasm is far from being shared by all perceptive thinkers. The composer Jacques Ibert wrote (ref. 20)

> Le mot système me fait horreur, et je fais le pied de nez aux règles préconçues. . . Tous les systèmes sont bons pourvu qu'on y mette de la musique,

which is, however, perhaps to say that a system which imposes its discipline implacably is stultifying, but that the artist can turn any system to advantage provided that he has the skill and creative power.

That methodology and analytical techniques in general are a blight on creativity has been frequently argued. Their potentially repressive influence is penetratingly exposed by Wylie Sypher in his book *Literature and Technology: The Alien Vision* (ref. 10). Sypher argues that technology, in contradistinction to science, necessarily constrains the initiative of the practitioner because he cannot afford to be wrong. He writes (ref. 10*a*)

> This industrial economy (one in which all non-profitable activity is eliminated) is based...like Mallarmé's poetic economy, upon an elimination of the accidental, or what the symbolists call Chance, *Hasard*, Prose. Existentialists call it the contingent, in contrast to the necessary, agreeing with Whitehead and other modern scientists that the most convincing evidence of reality is its uncertainty, the individual irregularity, leap or wilfulness that should not surprise us even if we cannot predict it. The engineer, that is to say the applied scientist, at his extreme must master the contingent. The technologist dreads surprises: indeed, he must not be surprised. He predicts everything – and in a sense, discovers nothing. Technism might be defined as a conquest without surprise – or perhaps a conclusion without risk.

Now it is true that a technologist designing an aeroplane or a bridge or even a bath tub can hardly face the possibility of its failure in use with equanimity. Indeed one has only to consider the thalidomide disaster to recognise how high is the price that may have to be paid by the unsuspecting public if the scientist permits himself to be surprised. But if modelling is used as a process in design, then it is possible for the designer to deliberately court surprise, deliberately to expose himself to the unexpected during the modelling stage when the penalty that has to be paid for ignorance, though not for stupidity, is small. Indeed this is a primary purpose of the activity and through this exploration of the fringe regions beyond those in which the prototype will be expected to operate, he seeks to protect all concerned against the potential catastrophe that might follow from his being surprised at a late stage in the project.

'The contingent' writes Sypher (ref. 10*b*) 'is a hazard which can no longer be ignored in art or science. Meanwhile the technologist continues to deal with the contingent by laws of probability, a calculus which rationalises the exception'. But surely we well understand that the unexpected is a hazard that can no longer be ignored when science is applied in ways which are likely to have significant social consequences. And it is very clear that the technologist must deal with the contingent, the unexpected, in his product by a calculus of probability that exposes and quantifies the risks to which he might be subjecting others, so that they may be reduced to an acceptable minimal figure. Sypher writes as though the applied scientist's task were to avoid facing up to uncertainty, in fact his task is to cope with it and to turn the unexpected to advantage.

Sypher (ref. 10*c*) quotes several times from a penetrating essay written in 1897 by Valéry

> Method calls for true mediocrity in the individual, or rather for greatness in the most elementary talents, such as patience and the ability to give attention to everything without preference or feeling. Finally the 'will to work'.
> . . .Practically, the difference between science and technology is defined by this criterion of limited initiative, (a criterion advanced by Valéry) which manifests itself in the programatic.

Valéry associated the modern consciousness of method with changes that occurred during the renaissance,

> and it is true that within renaissance art there was a fascination with method for the sake of method, that is, a mannerist tendency. Mannerism might be characterised as an art that sacrifices vision to method, and a considerable part of 19th Century art is mannerist in this way. Yet the renaissance exploiting of method in painting by Parmigianino or Bronzino, for instance, could achieve surprises because the method was still exploratory, a way of experimenting or reaching conclusions not entirely predictable. . .Furthermore the renaissance was a period of incoherent thought; the idea of a world order barely concealed an underlying sense of disorder. [Ref. 10*c*.]

and as we have seen in an earlier quotation Sypher draws a parallel between mannerism in art and the excessive pursuit of method in technology which he calls technism.

What special help can modelling techniques offer that might enable their users to escape from the methodological bondage that Sypher so clearly describes? Models can be used for significant experimentation with very little risk to those not directly concerned and they permit creative speculation during the learning stage. Models invite analogy and encourage lateral thinking. For example, Stafford Beer (ref. 15*b*) quotes an example of operations research in which a model of blockages building up in a steel rolling mill was constructed in the form appropriate to the reverberation of sound waves. This is not a pouring of new wine into old bottles, the process of analogical thinking can be a creative step. Models can enrich a problem by enabling large systems to be handled and by enabling the researcher to explore a variety of assumptions at many different levels quickly and effectively. They provide a description of intention that can serve as an excellent basis for wide discussion. But Sypher perceives a yet more sinister threat. He quotes (ref. 10*d*) Jacques Ellul

> inside the technical circle, the choice among methods, mechanism, organisations and formula is carried out automatically. Man is stripped of his choice, and he is satisfied. He accepts the situation when he sides with technique

and goes on to aver that, in an advanced society, decision can be remanded to the computer.

> Technicians can delude us into believing that we are free when we are not because techniques can absorb our very hostility to techniques because the technician can calculate our resentment in advance and provide for it in his programme. To leave us ignorant of being manipulated is the technician's supreme feat.

The problem is crucial, more so in politics and management than in the development of technology. The manager's 'supreme feat' should be to be so aware of his own motivation that he can enter into participative planning without attempting to 'con' the other participants. Can it be done? Or must we succumb to T. S. Eliot's greatest treason and do the right deed for the wrong reason (ref. 21)?

2.7.2 *A solution: Action research*

Concern about the social impact of the failure to recognise the fundamental openness of systems led Churchman in his book *Challenge to Reason* (ref. 22) to analyse the intractable nature of this problem to the point of near despair. His concern has been echoed elsewhere, for example Katz and Kahn (ref. 23) write

> A (second) error lies in the notion that irregularities in the functioning of the system due to environmental influences are error variances and should be treated accordingly. According to this conception, they should be controlled out of studies of organisations. From the organisation's own operations they should be excluded as irrelevant and guarded against. The decision of officers to omit a consideration of external factors or to guard against such influences in a defensive fashion, as if they would go away if ignored, is an instance of this type of thinking. So is the outmoded 'public be damned' attitude of businessmen towards the clientele upon whose support they depend. Open systems theory, on the other hand, would maintain that environmental influences are not a source of error variance but are integrally related to the functioning of the social system, and that we cannot understand a system without a constant study of the forces impinging upon it.

The systems approach is, fortunately, a very powerful instrument for ensuring that interactions with significantly high correlation coefficients are not ignored, and it can be fruitfully applied to problems in which social objectives and the manner of achieving them are essential components. One way of using the systems approach to integrating the social implications of change with the defining and solving of a problem is expounded in Checkland's paper *Towards a Systems-Based Methodology for Real-World Problem Solving* (ref. 24) which, additionally, provides a useful commentary on the origin and development of the systems movement.

Checkland is reporting the results of research in what he calls 'soft' systems. He writes

> In a previous paper [ref. 25] it has been suggested that the challenge facing those seeking to develop a systems

approach is to develop methodologies appropriate across
the spectrum from the relatively 'hard' systems involving
industrial plants characterised by easy-to-define objectives,
clearly defined decision-taking procedures and quantitative
measures of performance. . .(to). . .'soft' systems in which
objectives are hard to define, decision-taking is uncertain,
measures of performance are at best qualitative and human
behaviour is irrational.

and later

Similarly any concentration on definition of objectives
implies that an agreed definition of objectives for the
systems in the hierarchy will not be too difficult to obtain,
and will be useful once obtained; this is then the lead-in to
systems design: the design is such that these objectives may
be achieved. But in soft systems such definitions of
objectives may be impossible to obtain and if obtained may
not be useful. It may be that obtainable definitions are
'public' ones which conceal the 'real' objectives – perhaps
even from the people who profess them; or it may be that
objectives of a soft system are genuinely inexpressible; or
it may be, remembering Vickers' [ref. 26] distinction between
goal-seeking and relationship-maintaining, that it is simply
inappropriate to try to define objectives, objectives being
finite goals different in kind from the deeper underlying
ongoing purposes of a soft system which, though it may be
other things as well, will be a social system. And social
systems are not usefully regarded as goal-seeking.

Checkland's solution to the problem that he raises lies in 'action'
research, in which the researcher is himself involved in the process
of change in the system in which he is working and accepts that he
cannot necessarily design his research objectives into his experiments.
This development challenges the constrained view of the systems
method that requires a definition of objectives before research can
begin. It is a response to the inevitable contradiction between the
imperative need to act, an imperative dictated by the need to survive,
and the inevitable consequences of action in constraining the freedom
of others.

It might be argued that, in defence of modelling as a method, we are exemplifying the very fact to which Sypher and Ellul draw attention. And indeed it is certainly possible for models to be used in precisely the way that they deprecate. Many models used in social situations are virtually 'black boxes' that produce an answer to questions submitted, but through a process of reasoning which few who use them understand. And to this we shall return in the next paragraph. But as we have said in chapter 1, an important function of a model is to provide, for complex problems, that language and syntax which are essential if it is to become comprehensible, open, and vulnerable to criticism.

2.7.3 *The black box*

The term black box has frequently been used over recent years and in a variety of ways. We use it here to signify a model of which the inputs and outputs can be observed but of which the internal structure is unknown, that is to say that nothing is known about the way in which the box transforms inputs into outputs. This problem is particularly obvious in models that are run on a computer but it has general relevance.

For all but specialists, a computer will, in the last resort, inevitably be a black box, since even when it is used for simple computation and the rules for data input and of output are well understood, the structure of the operations within the machine are not. But the problem is not one of programming but of understanding the limitations that are inbuilt in available programmes.

Suppose that what we have is a computer programme for calculating the minimum safe thickness for a pressure vessel or for calculating the cash flows in a business operation. It has been written by people who had a clear perception of the nature of the problems with which they were dealing, but often the programmes come to be used by others who do not have this insight. If, however, the limitations inherent in the model are not explicit, if the user cannot readily appreciate how robust the model that he is using is, then he will not be able to assess how appropriate the model is to the particular task environment to which he wishes to apply it. The danger is serious because the use of computer programmes of this kind gives an air of conviction and finality to the results. It therefore behoves whoever writes such a programme to produce not only a User's Manual, which enables the user to go through the necessary operations mechanically, but also a description of the model that will enable the user to appreciate whether or not it is appropriate for the particular task environment to which he is addressing himself.

2.7.4 *The bromide – The substitute for action – The habit*

Many management decisions are taken in a situation of uncertainty in which adequate information is not available. Managers are often tempted to defer the moment of truth by calling for further evaluation, thus avoiding rather than coping with the problem. Moreover, it may be comforting if things turn out badly to say 'Well. I did every possible study; risk analysis, sensitivity analysis, modelling, and I had this and that programme written and run. What more could I have done?' It is always important to ask whether an exercise such as writing a computer model is likely to contribute significantly to uncertainty reduction or whether the time to decide has already come. It is often tempting to play with models as an alternative to tackling the actual problem. Talking about a problem has always been recognised as a substitute for action and the situation here is parallel.

Models can, like any operation, become part of the stereotype pattern of behaviour and it is very easy to say 'We have a model, let's run it every Monday'. Management information and procedures tend inevitably to grow fat. A positive effort to avoid running models unnecessarily is a useful contribution to keeping them slim.

2.7.5 *Better than toy trains*

Many people find that the intellectual discipline involved in programme writing and in the exercises that can now be done in conversational mode with computers are both stimulating and fascinating. One of the authors, who for a period developed an addiction to working with the computer, was effectively brought up sharply by one of his colleagues remarking that it was better than toy trains. Perhaps even this is all to the good, provided that time which ought to be spent in other activities is not spuriously devoted to the construction and operation of models that make little real contribution to problems that have to be tackled. Having said that we would emphasise that some familiarity with digital computers and programming is of enormous value not only in that it gives access to the machines and reduces the risk that one will be ever trapped in the black box situation, but also in that it provides a way of thinking, a way of structuring problems that can be usefully applied in very many situations.

We have now completed a review of the principles of modelling in which we have tried to illustrate the wide-ranging applicability of modelling concepts, not only in respect of problem variety but also in coping with problem complexity, and especially in maintaining relevance between the experimental or design situation and the task environment. In the chapters that follow we look more closely at the use of modelling techniques in situations that are of interest to the chemist and to the chemical engineer and to other scientists, especially biologists beginning with the structuring of thought in developing a basic understanding of chemistry and moving, through a consideration of complex molecules and biochemical problems, to industrial and commercial situations.

3

CONCEPTUAL MODELS IN CHEMISTRY

3.1 Introduction

Having approached models and their properties in a general and wide-ranging manner in the previous chapters we now turn to examine more closely the characteristics of models applied to specific situations that are especially relevant to chemists. The discussion builds up naturally through the following chapters beginning with atoms and small molecules and progressing through macromolecules and biochemistry to large scale production and innovation. We begin in this chapter by considering a model which lies behind every other model in chemistry, the chemist's personal conceptual model (subsequently referred to simply as the conceptual model) which, as the basis of the mental process of analysis, is his means of understanding a molecule or a reaction. This mental picture, the conceptual model, although containing many elements that are common to most chemists, is very much a personal thing. It is developed over the scientist's years as a student, influenced by his teachers and colleagues and by his personal experience.

Our discussion of the conceptual model begins with organic chemistry. This is because the variety of models used in organic chemistry is large, often disconcertingly so to chemists in other branches of the subject. Many of these models are matched by relatives in physical and inorganic chemistry and later in this chapter we examine models in these fields and compare them with those we have described from organic chemistry.

3.2 Classes of model

Let us firstly survey the organic chemist's conceptual models. They may be broadly classified as follows:

(1) Quantum mechanics
(2) Functional groups
(3) Effects
(4) Reacting systems

This classification offers a convenient framework for discussion and we shall see later how conceptual models appropriate to other fields of chemistry are complementary to these. All four classes of model interpret chemical behaviour in terms of certain defined structural units and ignore or minimise the relevance of other components of the system.

The *quantum mechanical model* is, in the terms used in chapter 2, the most theoretical and the most fundamental of the four, and the most comprehensive in this particular hierarchy (section 2.5.2). The structural units which it uses, essentially electrons and nuclei are, according to current theory, determining factors in defining most physical and chemical properties and in particular in the binding or repulsive forces between particles of matter. This is especially germane to chemistry, the science that deals above all with reorganisation of structural units in matter. In principle, all chemical properties, organic, inorganic or physical, are predictable from a knowledge of structure and the behaviour of the fundamental particles. But so complex are the possible interactions in most cases, that the necessary equations cannot be formulated precisely and on the rare occasions when they can, their solution is often impracticable, even with the aid of fast computers. This impasse may be circumvented either by using an exact mathematical model of a simplified system or a simplified model of the complete system. In either case we obtain a model in the form of equations that are soluble, at the cost of having a less precise representation of the prototype.

The second class of model, that of the *functional group*, is both more specific and more empirical than the quantum mechanical level. Organic chemistry is structured around the experimental recognition of similarities in the properties of compounds which contain the same reactive group of atoms – the well-known 'functional group'. The thinking is similar to that applied to elements in inorganic chemistry.

The importance of this class, as a point for entry into the conceptual model that does not require a theoretical base, has been emphasised by many authors. Three quotations, two from well-known organic chemistry texts and one from an advanced review, illustrate this point.

Classification according to functional groups has the virtue of providing a rationale for the behaviour of the vast numbers of organic compounds that are known. It is logical and convenient to divide organic compounds into families with similar properties in the same way that the periodic classification of the elements divides the elements into groups or families having related properties. [Ref. 1.]

One of the most important corner stones in the framework of contemporary chemical thought is the knowledge that reactions of compounds are largely determined by their functional groups. [Ref. 2.]

The importance of class three, effects, is also indicated.

Of equal importance, however, is the realisation that rate constants and positions of equilibria associated with the reactions of functional groups may be strongly dependent on whatever else is present in the reacting molecule.

The author here is making an oblique comment about the environmental interactions which we emphasise.

The structure, energy and chemical behaviour of organic molecules are all interrelated. The early chemist focussed attention on the functional groups present in the compounds they examined, but it quickly became apparent that wide variations were to be found in the rates and even in the course of reactions involving the same functional group. [Ref. 3.]

March (ref. 4) emphasises the usefulness of the reacting or functional group concept.

It is customary in writing the equation for a reaction of, say, carboxylic acids, to use the formula RCOOH, implying that all carboxylic acids undergo the reaction. Since most compounds with a given functional group do give more or less the same reactions the custom is useful. . .It enables a large number of individual reactions to be classified together and serves as an aid to both the memorisation and the understanding of them. Organic chemistry would be a huge morass of ill-related facts without the concept expressed by the symbol R,

and without the functional group level of conceptual modelling.

Of course, what groupings should be regarded as 'functional' depends very much on what type of reaction is under consideration. In an ionic reaction with a fatty acid, the nature of the hydrocarbon chain, except insofar as it affects solubility, may be of little or no importance and the symbol R comes into its own as a generalisation. On the other hand in a study of chlorination, when hydrogen abstraction is all important, CH_3- and $-CH_2-$ can no longer be regarded as equivalent. Functional groups are important because they can often be treated as quasi-isolates, but a large part of the organic chemist's skill lies in recognising and dealing with situations when this assumption breaks down significantly.

The fact that, to a large extent, the properties of organic molecules may be inferred from the functional groups present is symbolised by the use of such simplified formulae as RNH_2 for an amine, where R is a hydrocarbon group. The recognition that the behaviour of a functional group may be drastically changed by the nature of the R– to which it is bound was, no doubt, brought home to many readers for the first time when they learnt, for example, the differences between primary, secondary and tertiary and between aliphatic and aromatic amines.

Ideally, and the word is used both in its popular and thermodynamic denotations, the functional groups level of model requires that molecular behaviour be explicable in terms of the reactive groups alone, independently of their environment in the molecule of which they form part. We know that this is not so and the 'ideal' has to be corrected by application of more or less empirical factors – the *effects*. The model, at the effects level, brings a number of structural features of the molecule into the reckoning in addition to the functional groups.

Analogies based on class two models are highly successful in predicting qualitative behaviour but when reaction rates and equilibria are studied, quantitative differences become apparent that cannot be explained by the functional group analogies. Williams, Stang and Schleyer (ref. 3) comment that 'effects' were usually first introduced to explain quantitative differences in the observed properties of molecules of the same homologous series. For example, the inductive effect was developed to rationalise the widely differing acidities of substituted acetic acids (table 3.1, p. 74).

Effects may be no more than correction or 'fudge' factors that are introduced to convert a set of experimental data into a pattern that has an acceptable structure, in some cases fitting a predetermined mathematical expression. This alone would be useful, but it is perhaps in protest against arbitrary empiricism, which makes everything a special case, that Katritzky and Topsom comment (ref. 5) 'It is apparently a human failing, felt as much by chemists as by others, to try to visualise numerical relationships in terms of effects'. If Bruner is right in stressing the fundamental nature of enactive models then it is more a necessity than a failing. We suggest in chapter 2 that empirical models can be especially useful in expressing correlations in situations in which no extrapolation is contemplated. However, though this capability may be initially useful in clearing the way, it is hardly acceptable in fundamental research which aims at integrating new facts in the accepted corpus of scientific knowledge. To distort existing theoretical treatments in order to accommodate ill-fitting empirically based concepts is equally unacceptable as Gould (ref. 3.2) points out in defining the broad categories of effects. It is, of course, the task of the scientist to develop gradually a deeper understanding of his special cases so that they can eventually be described within the general terminology of science. In organic chemistry one would, fundamentally, look to thermodynamics or wave mechanics but the bridges have not yet been built.

Just as an elementary treatment of functional groups uses a symbolism which does not differentiate between different molecular environments, so also the simplest representation of a reaction system is silent on wider environmental issues, e.g.

$$A+B \rightarrow C+D,$$

in which reactants and products are symbolised but not reaction conditions. But we know that the phase in which reaction takes place, the nature of the solvent, the temperature, the incidence of radiation, the nature of surfaces present and so on are often decisive in determining the nature of chemical change. And so in the fourth class of model, the *reacting system*, the representation of this external environment is all important.

The chemist may work with a variety of conceptual models, those mentioned in this chapter and others as well, on the same problem, seeking insight from each class as and when seems most appropriate. There is, however, much to be said for considering molecules or systems of reacting molecules as the basic descriptive units whenever possible. The actual molecules are then the prototypes whatever representation we choose to use as the model.

Let us summarise the argument so far.

Organic chemistry provides an excellent example of the growth of a complex conceptual model through an alternation of observation and empirical modelling with subsequent working out and restructuring of the model in theoretical terms. The discovery of families of compounds with similar properties was systematised by the development of concepts such as those of the homologous series and of functional groups. The pattern of properties displayed by each group was interpreted in terms of its structure and the nature of its constituent elements.

For example, an alkyl halide contains a halogen atom that can be readily substituted by nucleophiles,

$$CH_3-Cl + N(CH_3)_3 \rightarrow CH_3-N(CH_3)_3^+Cl^-,$$

but the halogen atom in an aryl halide, C_6H_5Cl, is totally inert under these conditions. The difference can be explained by an interaction between the p orbitals of the halide and the π orbitals of the aromatic ring which causes a stronger carbon–halogen bond to be formed than in the alkyl case in which π orbitals are absent. This explanation embodies theoretical concepts additional to those that were necessary for the generation of the idea of a functional group.

Experiment then showed that the behaviour of accepted functional groups, especially when quantitative data rather than qualitative results were considered, depend on the nature of the remainder of the molecule. These experimental differences were systematised in an empirical model that introduced 'effects' to explain them. But, as always, scientists felt uncomfortable with a model that contained empirical parameters that were not related to the wider theoretical model and, as we shall see, effects were 'explained' in theoretical terms and the validity of the explanation was tested, in due course, by experiment. The process was similar when the focus was deepened from the single molecule to the reacting system. At each stage the scope of the concepts in play was extended by applying them by analogy to new situations. Discrepancies between the expected and the observed demonstrated the inadequacy of the current model.

Each stage of the enlargement of a model brings previously ignored environmental factors into account. This may occur without shifting the interface between inner and outer environment, as when, with a molecule as prototype, we introduce the concept of effects. When, however, we move to the study of reacting systems we could take the whole system as prototype, thus shifting the inner–outer interface beyond the individual molecule, or we could choose to think still in terms of a molecule as prototype with the rest of the system now consciously considered as the environment. As we suggested earlier, the latter is usually more convenient.

We now turn to a discussion of each of the four levels of model entering at the easiest point, the concept of functional group. Many of the molecules we shall discuss in later chapters are large and contain several types of functional group. It is an essential part of the skill of an organic chemist to interpret the reactivities of these groups and in the next section we discuss an example that requires this skill.

3.2.1 *Functional groups*

As an example of a complex molecule with an array of functional groups we have chosen terramycin, an antibiotic of the tetracycline group (figure 3.1).

The tetracyclines take their name from their common structural feature – the four linearly fused six-membered rings. One of these is

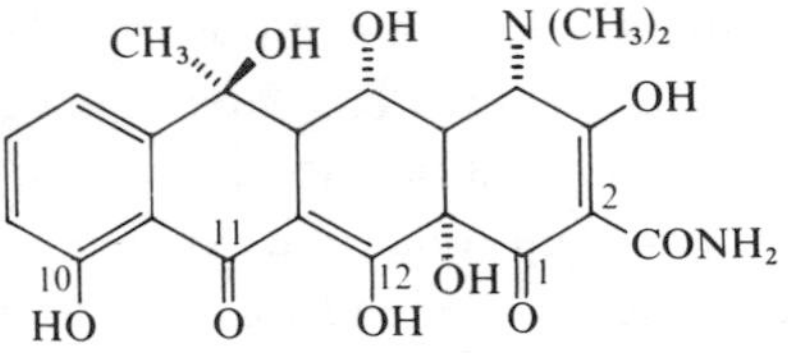

Terramycin

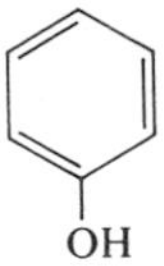

(3.1.1)

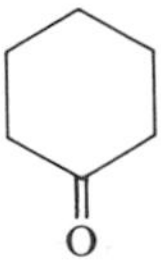

(3.1.2)

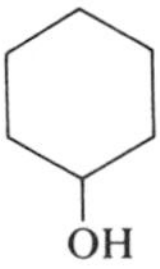

(3.1.3)

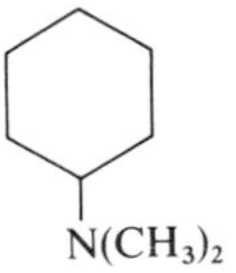

(3.1.4)

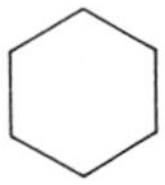

(3.1.5)

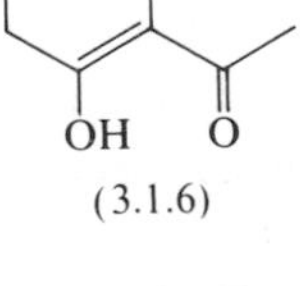

(3.1.6)

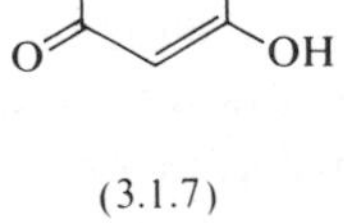

(3.1.7)

Figure 3.1. Terramycin and model compounds.

aromatic and the other three are more saturated. Disposed around these rings is a wide variety of functional groups, the behaviour of which we can compare with the same functional groups in simpler compounds. These simpler compounds may therefore be considered as *model compounds* for terramycin (3.1.1–3.1.7) in respect of those characteristics that depend upon the particular functional group. Each model compound contains only one functional group but it is clearly possible that the functional groups in the prototype terramycin are capable of interacting with each other. Such isolated functional groups are standard systems in the conceptual model and so we can compare their behaviour with like groups in more complex molecules. Most commonly we may wish to predict the properties of these molecules and to rationalise experimental results. We must bear in mind that functional groups and model compounds are only quasi-independent because of potential interactions between groups.

By way of an example let us attempt to describe the properties of terramycin as far as possible using the conceptual model. On examining the structure we immediately identify the various functionalities present – a phenol, a secondary alcohol, a tertiary amine, etc. Then we can attempt to predict the chemistry of each functionality in turn. As we are operating within class two, functional group models, we can neglect any interactions which occur between these groups for the time being.

Consider the phenol group at C-10. How are we to predict its properties? The answer is to use an analogy which relates our phenol group to a model compound containing a hydroxyl group of known properties, for example phenol itself (3.1.1).

One of the first things a chemistry student learns is a network of such analogies, and accordingly he is taught the chemistry of one or two representative members of each class of functional group. Other possible analogies, or models, for the other functionalities in terramycin are illustrated by the structures (3.1.2–3.1.7).

Some parts of the molecule may be ignored. For example, it is essential to this process of analogy that the alkyl rings or chains on which the functional groups are suspended can be regarded as chemically inert (e.g. 3.1.5). Some interactions, however, cannot be ignored. Cyclohexanone (3.1.2) would not be a good model for the carbonyl group at C-11. This is because the carbonyl group at C-11 is conjugated with the enol form of another carbonyl group at C-12. A better analogy would be the compound (3.1.6) or (3.1.7) which also contains the conjugated system of double bonds. As with all examples of modelling by analogy it is essential that we have a good analogy, that the model–prototype transformation is as reliable as possible.

On the basis of such comparisons we could predict, amongst other things, that terramycin would form a phenolate anion at pH 9 just as a phenol does and that it will form chelate complexes through the diketone functions analogous to the behaviour of compounds like (3.1.6). Indeed this latter property has been suggested to be the chemical basis of the antibiotic activity of terramycin.

In this section we have seen how it is possible to make reasonable predictions of the chemistry of an unknown compound by a process of analogy using class two models.

These are the first models to be called upon because they respond most readily to the usual drawing of a chemical structure (see section 4.3). Let us recall however one of the quotations given earlier (ref. 3). 'Wide variations were to be found in the rates and even in the course of reactions involving molecules with the same functional group.' The cause of, or the rationalisation for, these variations lies in the members of the third class, the effects.

3.2.2 *Effects*

The functional groups provide a qualitative framework upon which models of reacting systems can be built but they can only suggest the type of reaction that may be expected. Reactivity, observed as differing rates of reaction or of equilibrium position, cannot be predicted. Without further aid we are left to consider functional groups in isolation, as independent systems. However experimental evidence leads us to expect that functional groups in polyfunctional compounds are capable of interacting with each other. The models of class three (effects) attempt to give a physical basis to these interactions and to quantify them by measuring the magnitude of the interaction of one functional group with another in the same molecule.

Clearly then, models of class two and class three are both confined to the internal environment of the molecule. All other influences that a molecule experiences are due to the external environment in which it finds itself. Such external effects may arise from solvent, temperature, and from other reactive molecules amongst many other things, and these considerations bring us to the fourth class of models (reacting systems) which we discuss in the following chapter.

The most common effects. There is a strong subjective element in the analysis of the interactions between functional groups. This is because diverse phenomena that may be interpreted from experimentation as discrete, may be fundamentally caused by the same interaction of functional groups. Nonetheless effects are mainly described in empirical terms: they are plain rationalisations of experimental data though they cannot always escape from looking like theories that have been stretched beyond their elastic limits in order to bring them into closer correspondence with reality.

It is, however, generally agreed that groups interact by any of three major mechanisms known as inductive, resonance or mesomeric, and steric effects. This division is described by Gould (ref. 2) as, 'although not based on thermodynamics, by and large congenial with the thinking of organic chemists'. To tie these effects down closer to a physical picture of molecules, we can associate an inductive effect with the polarisation of a bond caused by an attraction of the bonding electrons towards the more electronegative partner in the bond. A resonance or mesomeric effect is the interaction of groups or atoms through a system of conjugated multiple bonds, formally interpreted as the movement of electron pairs. Inductive and resonance effects together describe what happens to the electrons in a compound when its functional groups interact and they can therefore be grouped together as the electronic effects within a molecule. Steric effects can be more easily visualised. They simply represent interactions due to the bulk or shape of an atom or group: such interactions commonly occur when one group 'gets in the way' of another.

Quantifying effects. These descriptions of effects give us little basis for a quantitative understanding of the interactions between functional groups in molecules that are a cause of variations in reactivity, but they do provide a useful conceptual framework. If we are to estimate the magnitudes of effects such as these, then we must find a system in which only one effect can reasonably be the cause of the reactivity differences, and secondly, we must define a closely related reference system in which that effect does not operate at all.

As an illustration, let us suppose that we wish to find reasons for the differences in acidity (pK_a) of substituted acetic acids (table 3.1). The first condition, absence of multiple effects, is easily met. Since there are no conjugated double bonds, no resonance effect exists and the substituents that we are varying are remote from the acidic CO_2H

group. Therefore no steric effect operates either. It is equally easy in this case to define the reference system – acetic acid itself in aqueous solution. The electronegativities of carbon and hydrogen are similar and the C–H bond is accordingly scarcely polarised at all. Hence it is reasonable to take an isolated C–H bond as exerting zero inductive effect. We can therefore understand the observed increase in acidity that accompanies the substitution by more electronegative elements for hydrogen as a consequence of the inductive effect in the C–X bond. The polarisation is transmitted by further induction through the C–C and C–O bonds. Thus the charge density present on oxygen in the anion is reduced. Delocalisation of charge is a stabilising influence on the anion in a polar solvent like water and consequently the greater the withdrawal of electrons, the stronger the acid.

Table 3.1

Acid	pK_a
$CH_3–CO_2H$	4.75
$Cl–CH_2–CO_2H$	2.86
$Cl_2–CH–CO_2H$	1.80
$Cl_3–C–CO_2H$	0.65

In this case, it is easy to marry our pictorial concepts to hard experimental facts because both the reference system and the effects operating are clear-cut. Things are rarely so simple, and difficulties often arise from the definition of the reference system. Since the reference system is a model of that specific conceptual situation in which an effect is absent, a change in the reference system can imply a change in the interpretation of the experimental data. Consequently, controversy can easily arise about the validity of a reference system in relation to a particular effect.

An argument about reference systems. One such controversy was the debate over the existence of 'non-classical' carbonium ions* as inter-mediates in certain solvolysis reactions. Figure 3.2 shows a general-

* The general term for positively charged carbon species is carbocation. Modern nomenclature refers to pentacoordinate positively charged ions as carbonium ions. This group includes the non-classical ions discussed in this section. Triply coordinate species are referred to as carbenium ions.

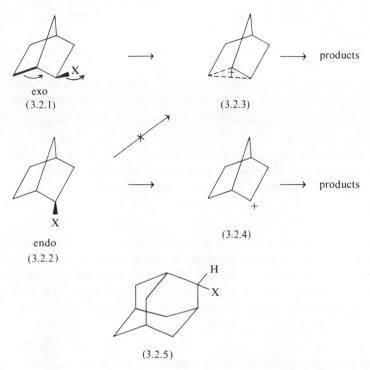

Figure 3.2. 'Non-classical' carbonium ions.

ised case in which the substituent, X, is a good leaving group such as p-toluene sulphonate (ref. 6). Experimentally it was observed that the rate of solvolysis of the isomer in which X occupies the *exo* configuration (3.2.1) was 340 times faster than the rate of solvolysis of the epimeric *endo* isomer (3.2.2). As we shall see in the next chapter, the higher solvolysis rate of the *exo* epimer implies that a lower energy pathway exists for its solvolysis than for the *endo* isomer. It was argued that the *endo* isomer underwent solvolysis via a more or less normal S_N1 mechanism (chapter 4) through a transition state in which positive charge tends to build up on only one carbon atom (3.2.4). On the other hand, the *exo* isomer can solvolyse via a more stable transition state. As X moves away, the C–C bond coplanar with C–X joins in the reaction by donating a share of its electrons to reduce the developing positive charge. Eventually, a delocalised or 'non-classical' ion results (3.2.3) that is of lower energy than the localised classical ion (3.2.4). This is sometimes referred to as a bridged ion. In the *endo* case, there exists no such coplanar C–C bond and therefore stabilisation of the transition state by charge delocalisation is impossible.

The controversy raged over how great a contribution, if any, such C–C bond participation (referred to as anchimeric assistance) made to the difference in solvolysis rates. In order to assess the contribution from a non-classical ion it is necessary to establish a reference system in which anchimeric assistance cannot occur. Here was the difficulty. The *endo* isomer won't do because the C–X bond has the wrong configuration and anyway, it is already part of the argument. The protagonists of anchimeric assistance were able to bring forward reactions in which they could confidently rule out its operation. However their opponents countered with the argument that if these reactions are essentially of the same nature as the disputed reaction and if they can be explained without postulating non-classical ions as intermediates, then there is no justification for invoking non-classical ions at all. It has since been demonstrated (ref. 7) that delocalised carbonium ions do exist under very special circumstances but even this result cannot be accepted as conclusive evidence for the intermediacy of delocalised ions in solvolyses because the conditions are so different.

The whole story is an example of the growth and development of the conceptual model from experimental results. One group of workers feel that an extension of the model in the form of a new phenomenon (anchimeric assistance) and a new intermediate is required by experimental results. Their opponents assert that the conceptual model as it stands is capable of interpreting the results. Both base their arguments upon selected pieces of information, the subjective judgment that is inherent in all controversy. In this case, it was difficult to prove conclusively that a new effect was absent and, reviewing the evidence, Bethell and Gold comment (ref. 8):

> Estimation of the extent of anchimeric assistance directly from solvolysis rates is difficult since the values depend markedly upon the reference compound... In practice the difficulty of using the rate increase as a criterion of bridged ion formation lies in predicting the reaction rate in the absence of assistance by the bridging group.

Not all reference standards have been the subject of such controversy, fortunately. The substituted adamantane (3.2.5) is another example of a reference compound chosen to limit the argument to one effect only. Again X represents a suitable leaving group. The study concerned the relative rates of solvolysis of secondary and tertiary alkyl halides and it was desired to have a reference compound which excluded the possibility that a solvent molecule might approach the rear of the reacting centre. A cage structure like adamantane makes such an approach sterically impossible. Thus by standardising the interaction of solvent, we are controlling the outer environment of the reacting system and concentrating entirely upon the interactions of the desired reactant molecules (ref. 9). Recently Brown has provided an extensive and readable discussion of the non-classical ion question.

These are some of the problems that are encountered when we try to compare effects within a closely related series of reactions. If the effects are transmitted intramolecularly, it should be possible to devise a measure of the ability of a given group of atoms to give rise to a particular effect. In principle, quantum mechanical models should be capable of achieving this but the first useful correlation of groups with the magnitude of the effects that they produce (the Hammett equation) was derived empirically.

Linear free energy relationships. The Hammett equation. Hammett was the first to relate an experimentally observed effect quantitatively to a parameter characteristic of the group causing it. Choosing reactions of a series of compounds for which much accurate experimental data was available namely *meta* and *para* substituted benzoic acids, he perceived that a regular correlation held (ref. 10). On plotting the logarithm of the equilibrium constants for ionisation of *meta* or *para* substituted benzoic acids against the logarithm of the corresponding rate or equilibrium constants for another reaction of the acids (esterification, for example), a straight line was obtained. The procedure involves the two reference systems which we recognised for the previous examples: the unsubstituted compound, benzoic acid itself forms one reference point, and a reaction that occurs without the immediate intervention of another reactant, the ionisation of the series of substituted benzoic acids in water, is the other. A typical plot is shown in figure 3.3. The linear correlation observed for the ionisation of the two series of acids, benzoic and phenylacetic, can be expressed as an equation of the form

$$\log K = \rho \log K' + C$$

where K' is the equilibrium constant for benzoic acids, and K that for phenylacetic acids. We can see that this holds for any *meta* or *para*

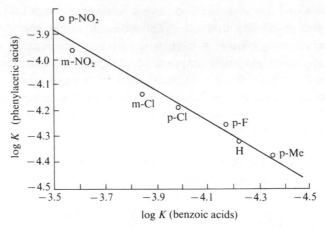

Figure 3.3. A correlation of substituent effects.

substituent including our reference hydrogen. Therefore an analogous equation can be written for the reference compounds

$$\log K_0 = \rho \log K_0' + C.$$

Subtracting these equations we obtain

$$\log \frac{K}{K_0} = \rho \log \frac{K'}{K_0'}.$$

This would hold for any reactions in which a linear log–log plot of rate or equilibrium constants is obtained. It can be simplified if the term $\log K'/K_0'$ is defined as σ, a constant characteristic of a given substituent and this term can be readily evaluated from the available accurate experimental data for substituted benzoic acids. The equation is therefore reduced to

$$\log \frac{K}{K_0} = \rho\sigma,$$

and it is known as the Hammett equation.

The substituent constant, σ, provides a measure of the electronic interaction of a substituent with the functional group at which the reaction takes place, ρ is referred to as the reaction constant and it is a measure of the sensitivity of the reaction to the effect of a substituent. Many reactions exhibit good linear correlations of log K with Hammett

σ constants, for example, the hydrolysis of benzoate esters fits the Hammett equation giving a value of 2.54 for ρ. However *ortho*-substituted benzoic acids do not fit this pattern of behaviour. In this case, the relationship between the reference system and the subject of the study must have altered, and another effect apart from the electronic interactions has come into play. We can assign the difference in behaviour to a steric effect in which the *ortho* substituent imposes a change in the surroundings of the site of the reaction.

It is important to appreciate how the reference systems are models of certain complex patterns of chemical behaviour. Even if we cannot describe precisely what is behind these patterns of behaviour, we must be able to recognise when something does not fit. This was why Hammett chose the system with the most accurate and extensive experimental data upon which to base a correlation. It has subsequently been found that the general approach allows one to compare substituent interactions in situations as different as mass spectrometry and enzyme kinetics (for example ref. 11) and perhaps more importantly, to spot the unusual.

Chemically speaking, although we can quantify and integrate a series of experimental results into a tidy pattern using Hammett's approach, we are still left with a strong element of empiricism because it is not possible to say from the correlation alone how the effect observed is mediated. That is, it is hard to separate the contributions of inductive, resonance, and steric effects quantitatively. Westheimer has commented, 'These effects often interact. If a quantum mechanical problem connected with any particular reaction could be accurately solved, a division of chemical effects into three or more categories would be unnecessary' (ref. 12). In other words a set of semi-empirical models would be subsumed in and therefore explained by a model at a high level in the hierarchy. The division, as Gould commented, is one of convenience and allows functional group interactions to be understood using a simple conceptual model. The separability of effects is discussed in detail in a review (ref. 13). Since free energies of reactions are related to equilibrium constants by the equation

$$\Delta G = -RT \ln K$$

correlations such as the Hammett equation are often referred to under the generic title of *linear free energy relationships* and we shall meet them again in the next chapter. For the meantime, let us turn to look more closely at steric effects and the concept of strain in molecules.

Steric effects and strain. The analogy between a chemical bond and
a stretched spring is a natural one and the concept of strain arises
directly from it. Baeyer in the nineteenth century realised that high
reactivity was often associated with strained structures. Today we
understand strain as the stretching, twisting, or compressing of bonds
out of the most stable configuration that is normal in the absence of
strain. The simplest molecules are devoid of strain and to describe
strain in complex molecules demands a very sophisticated quantum
mechanical treatment. A more amenable approach is to consider strain
as the potential energy stored within a molecule by classical mechanical forces. This in turn leads to an empirical quantitative model
that can be developed by factorising strain into a number of terms each
of which relates to a different sub-effect. These include any abnormal
bond angle or bond length and any steric interaction between groups
not chemically bonded together. Calculations of strain are frequently
based upon a factorisation due to Westheimer (ref. 12),

$$\text{strain energy} = E_{\text{bond angle}} + E_{\text{bond length}} + E_{\text{torsional}} + E_{\text{non-bonded}}.$$

Each term can be evaluated independently using methods derived from
classical mechanics. The use of the word 'abnormal' above implies
that once again it is necessary to define suitable 'normal' reference
systems in which strain is absent. Apart from the classical mechanical
models that this equation introduces, it is, of course, itself an approximation because in practice, an exact partitioning of the molecular
deformations into four terms is not possible.

As we have just mentioned, a series of reference compounds must
be defined if we are to make use of Westheimer's factorisation. It is
much harder to do this for steric effects, so much so that Williams,
Stang and Schleyer (ref. 3) assert: 'Any choice of strain free model must
be arbitrary...The strain free lengths and angles needed to evaluate
[Westheimer's factorisation] are really no more than adjustable parameters meaningful only in terms of the calculation in which they are
used'. The situation is not as bleak as it sounds, however, because if
the functions in the equation are calibrated carefully using accurate
experimental data from well-chosen reference compounds, then calculations can be made which agree well with other experimental data.
Workers in the field are well aware of the limitations of the approach,
'Applications of this method are generally successful when applied
to compounds whose structures vary but little from those for which

the parameters were evaluated' (ref. 3). This is what we would expect and it applies equally to a wide range of semi-empirical models including the linear free energy relationships.

The validity of empirical models of effects. If empirical models are successful, then there is some justification for the simplifications implicit in their design. The simplifications may, however, be chosen for convenience alone and not relate to an underlying physical or chemical reality. This distinguishes semi-empirical approaches from more fundamental procedures. The semi-empirical model is chiefly a tool to be used to obtain useful predictive results. Nevertheless, it is often necessary for a model of a chemical phenomenon to evolve from a descriptive and conceptual ancestor through an empirical but quantitative offspring to grow into a mature theoretical model.

Practical chemistry contains so many variables that an effect is a common qualitative stand-by model. If experimental results do not match the predictions of a current model, chemists rationalise the situation by proposing that a new effect operates. This can lead to a proliferation of effects many of which obscure both their origin and significance, an undesirable state of affairs. As Schleyer put it: 'There is an unfortunate tendency in the interpretation of organic phenomena to invent ad hoc explanations for each fact not conforming to some pattern of expectation. A corollary to this tendency, no less unfortunate, is the over-rating of an effect' (ref. 14).

3.2.3 *The theory of hard and soft acids and bases*

In the previous sections we have attempted to describe chemical reactivity using models of classes two and three. Many of the models used depend at least in part on empirical correlations which work well for the small number of first row elements common in organic chemistry. However when heavier, or highly electropositive elements are involved, conceptual models of effects and functional groups are less used. Chemists have sought to tie their knowledge to other frameworks that will more readily accomodate inorganic as well as organic reactions.

Such a classification has been attempted by Pearson (ref. 15) and he calls his model of chemical reactivity the theory of hard and soft acids and bases (abbreviated HSAB). Pearson divides functional groups and other reacting species (including inorganic ions and small molecules) into two classes, acids and bases, in which the terms are used in the

Brønsted or Lewis sense. The members of the classes of acids and bases are then divided into two groups, known as hard and soft. It is important to realise that hardness and softness are empirical properties and merely reflect the observed situation that a hard acid tends to react more readily with a hard base and a soft acid with a soft base.

Hardness and softness can be approximately understood using the conceptual model we have just described: hardness is related to a low polarisability and high charge density of an atom or functional group and softness to high polarisability and low charge density (ref. 15). Thus anions such as O^{2-}, Co^{2+}, Na^+, $CH_3COCH_2^-$, and BH_4^- are considered hard but I^-, Pb^{2+}, and $(C_6H_5)_3C^+$ are regarded as soft. There is, however, no need to relate the HSAB classification to any other model. The HSAB analysis at present is purely empirical and it has not yet proved possible to quantify hardness and softness, although attempts to relate these terms to quantities which can be calculated by quantum mechanics have been made (ref. 16). A short example will illustrate how this model operates.

An alkyl sulphenyl iodide, RSI, is a stable compound whereas the corresponding fluoride, RSF, is unstable. The HSAB method rationalises this experimental observation by considering the following hypothetical reaction.

$$RS^+ + HX \rightarrow RSX + H^+.$$

The sulphonium ion RS^+ bears only one unit charge on a weakly electronegative sulphur atom which has in addition two lone pairs of electrons not tightly held. Such properties are associated with high polarisability and are consistent with the sulphonium ions being soft. Similarly iodide ion, a large anion bearing a single negative charge, is also soft. On the other hand, a small ion of an electronegative element would be hard, a typical example being the fluoride ion here. Therefore the soft base iodide would be expected to form the more stable compound with the soft acid RS^+ and this is what is observed.

3.2.4 *Quantum mechanics*

We have delayed discussing what is perhaps the most fundamental description of matter, quantum mechanics, and its relation to the conceptual model. This is because we have approached the conceptual model from the point of view of the organic chemist and for him the enactive and iconic models of other classes are more often part of day-to-day thinking. In other words he will tend to think more in pictorial terms of the orbitals taking part in a reaction rather than considering the basic mathematical expression, the wave function, from which the orbitals are derived.

In this section we attempt to redress the balance by discussing some elementary modelling procedures found in class one quantum mechanical models. Then, in the following two sections, we broaden the discussion to the fields of physical and inorganic chemistry.

Wave functions, orbitals and atoms. The very word orbital can imply many different things depending on the matter to hand. At the most refined an orbital can be an expression of a wave function which is an exact solution of the famous Schrödinger equation. The Schrödinger equation (equation 3.1) has only been solved for simple chemical systems such as the hydrogen atom or the hydrogen molecule ion (ref. 17).

$$E\Psi = H\Psi.\tag{3.1}$$

H is known as the Hamiltonian operator and describes the total energy of the system in question. It has the character of a complex differential operator. E is the energy of the electron that is described by the wave function. The wave function contains an exponential and a polynomial term which correspond respectively to a radial and an angular function when polar co-ordinates are used (equation 3.2).

$$\Psi = R_{nl}(r)\,Y_{lm}(\theta,\,\phi).\tag{3.2}$$

(n, l, and m are integers known as quantum numbers.)

For the hydrogen atom typical solutions of the wave function are equations 3.3 and 3.4 which give the shape of the 1s and the 2p$_x$ orbitals in terms of the polar co-ordinates.

$$\Psi(1s) = \frac{1}{\sqrt{\pi}}e^{-r}.\tag{3.3}$$

$$\Psi(2p_x) = \frac{(2-r)}{2\sqrt{2}}e^{-r}\frac{\sqrt{3}}{2\sqrt{\pi}}\cos\theta\sin\phi.\tag{3.4}$$

In this case everything, including the energy levels given by E, can be calculated precisely but the mathematical solution of the differential equations which apply to polyatomic or polyelectronic systems are too complicated to be soluble and progress requires that simplifying assumptions are made. A model is constructed. At the most simple and easily visualised level the orbital is considered as a charge cloud which the rapidly moving electron produces. These clouds, familiar as stretched squashed and knotted sausage shapes, are abstractions of graphical representations of the wave functions or orbitals (refs. 17, 18a) (figure 3.4). Such representations are further abstracted by conventional chemical structural formulae which we shall discuss in more detail in chapter 4.

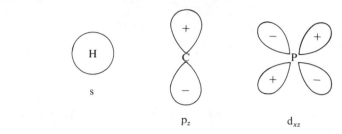

Schematic drawings of orbitals

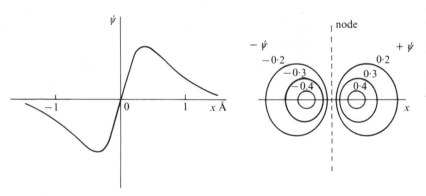

Figure 3.4

Between the two levels of abstraction, orbitals may be represented as precise graphs or as a table of computed data. The choice of presentation depends simply on the nature of the problem to hand. An atomic spectroscopist, for example, would wish to describe the excited states of atoms very precisely in theoretical terms because he can measure spectral lines very accurately. On the other hand a synthetic chemist, interested chiefly in obtaining a few milligrams of end product, will be satisfied if considerations of the interactions of the sausage shaped orbitals suggest likely reactions. In fact one can make considerable progress with theoretical considerations in both ways.

As we have already commented, the complexity of quantum mechanics arises from the impossibility of solving the Schrödinger equation for many-electron systems. However a solution has been achieved for the hydrogen atom and an obvious and legitimate modelling procedure would be to base wave functions and orbitals for more complex systems upon the hydrogen atom solutions as prototypes and to tailor the results as necessary to obtain the best agreement with experiment. This modelling can be approached in two ways, either by obtaining an approximate solution to an exact Schrödinger equation by mathematical methods such as an iterative optimisation procedure to give a minimum energy of the orbital, or, more commonly, by computing an exact solution from an approximate Schrödinger equation.

Even further simplification is required for calculations relating to larger molecules. A common assumption in these cases is that the nuclei of the molecule, being of very much greater mass than an electron, can be regarded as remaining stationary whilst the electrons move about them (the Born–Oppenheimer approximation ref. 18b). Within the framework of the Born–Oppenheimer model many different descriptions of molecules have been built up and we shall look briefly at the characteristics of the two best known. A complete description within a few paragraphs is not possible and thorough and detailed treatments can be found in most modern books on valence theory (e.g. refs. 18c, 19a and 20).

Orbitals and molecules. In each case the starting point is the atomic orbitals appropriate to the atoms in the molecule of interest. These orbitals are derivatives of hydrogen atomic orbitals. For example, in organic calculations orbital functions known as Slater orbitals are often used. These are a first-order approximation of the Hartree–Fock atomic orbitals (equation 3.7) and are derived by taking only one term of the summation (ref. 19b). Slater orbitals for carbon:

$$\Psi(1s) = 7.66e^{-10.8\,r}. \tag{3.5}$$

$$\Psi(2p_x) = 3.58r\cos\theta e^{-3.07\,r}. \tag{3.6}$$

The atomic Hartree–Fock orbital (ref. 18d):

$$\Psi = \left(\sum_{n,\,\zeta} c_{n\zeta}\, r^n e^{-\zeta r} \right) Y_{l,\,m}(\theta,\,\phi) \tag{3.7}$$

n, l, and m are quantum numbers. c and ζ are constants appropriate for each term and orbital respectively. The latter orbitals are most useful in the study of polyelectron atoms.

By now the reader will appreciate that the term orbital is used to refer to any of the expressions of the wave function from algebraic to diagrammatic. Also it is clear that our simplified orbitals are still valid models of the known hydrogen orbitals because they have the same mathematical form.

Valence bond theory. To construct a quantum mechanical model of a molecule, the atomic orbitals of each atom are combined to produce wave functions referring to the molecule and there are different ways in which this may be done. The first successful method, the *valence bond* or VB theory as it became known, was initiated by Heitler and London in 1927 (refs. 19c, 21). They described the molecule as if two electrons each from two different atoms formed a covalent bond in such a manner that the two electrons could never be found simultaneously on the same atom, that is they move together in a completely correlated manner. A wave function for the molecule was then constructed to embody this model. For the hydrogen molecule, for example, we can use the hydrogen atomic orbitals which are solutions of the Schrödinger equation. Thus for atoms A and B with electrons of opposite spin (1 and 2) we obtain the molecular wave function of equation (3.8).

$$\Psi = \phi_A(1)\,\phi_B(2) + \phi_A(2)\,\phi_B(1) \tag{3.8}$$

where ϕ_A, ϕ_B are the hydrogen atomic orbitals and $\phi_A(1)$ refers to the wave function when electron 1 is associated with atom A.

Evaluation of the bond energy of H_2 using this model, which is

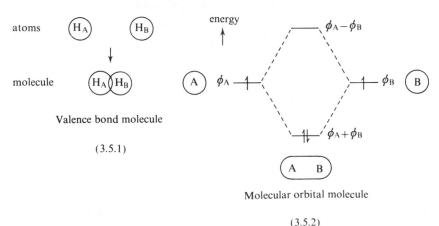

Figure 3.5. VB and MO treatments of the hydrogen molecule.

clearly a substantial simplification, yields a value of about two-thirds of that derived experimentally.

The VB model found ready acceptance because it embodies the familiar concept of the electron pair bond which forms the basis of most conventional structural formulae. In a pictorial sense this may be described as the overlapping of two atomic orbitals as shown in (3.5.1).

Molecular orbital theory. An alternative and subsequently more highly-developed method is the *molecular orbital* or MO theory introduced at about the same time as the VB approach by Hund and Mulliken (refs. 19d, 22). Here, rather than considering the atoms as interacting independently with each electron, wave functions are built to encompass the whole molecule as a unit and each wave function can accommodate two electrons of opposite spin. One simple approach is to construct molecular orbitals (MOs) by taking linear combinations of atomic orbitals (AOs). For the hydrogen molecule we obtain two MOs; equations 3.9 and 3.10.

$$\Psi_1 = \phi_A + \phi_B. \tag{3.9}$$
$$\Psi_2 = \phi_A - \phi_B. \tag{3.10}$$

These MOs can be represented on an energy diagram which indicates that a more stable system is formed if two electrons are placed in the lowest lying MO. The diagram (3.5.2) and the pictorial representation of the MO can be considered as another model of the covalent bond in H_2.

With class one models the only check we have that our prototype–model transforms are sound is comparison of the predictions of the model with experimental data. Discrepancies must lead to modification or total rejection of the model. This is the price paid for the intrinsically greater rigour of a quantitative calculation over a qualitative description. Physical chemical models (section 3.3) and theories of reacting systems (chapter 4) show precisely this feature but in distinct ways. Quantifications of effects (section 3.2.2) are comparative rather than deductive and hence are merely refinements of models within class three.

Much higher accuracy in calculations of the energy of the hydrogen atom can be obtained using these two methods as basis. In such simple binuclear molecules the differences between the VB and MO theories are not as apparent as they are in descriptions of larger molecules. For example, VB theory would consider cyclopropane to be essentially the sum of three separate strained carbon–carbon bonds all of which are identical. In contrast MO theory would describe the bonding as being due to three carbon–carbon MOs which could be represented pictorially as in figure 3.6. In each such orbital all three carbon atoms participate. Several different quantum mechanical models of cyclopropane can be constructed depending upon how the linear combinations of the orbitals are chosen (ref. 23).

Figure 3.6. Molecular orbitals of cyclopropane.

The MO approach has turned out to be the more adaptable of the two approaches that we have considered here and applications of MO theory are found in all fields of chemistry from organometallic compounds (section 3.4) to the Woodward–Hoffman rules (section 4.7). From a modeller's point of view we might expect the MO method to be more successful in describing molecular properties such as electronic spectra. Similarly when the important feature of a molecule is localised upon a single atom such as in a diradical the VB structure has more to offer. The criterion by which such models are judged is solely their agreement with experiment and it is futile to argue about the relative complexities or the aesthetic appeal of the different approaches in order to decide which is best applied to a problem.

Quantum mechanics is a no-man's-land between chemistry, physics and mathematics and to probe further would attract the cross-fire of specialists. Even at this elementary level we can see how the fundamental assumptions in a model define the validity of its results. Pople has made this generalisation in similar terms (ref. 24).

A more modest, but realistic approach is to adopt several clearly defined levels of approximation which do not represent the ultimate possible for small molecules but are simple enough to be widely applied. Each can then be tested against 'real' chemistry where experimental data are available and, if there is consistent success, some confidence is gained in its predictive power.

In this passage Pople summarises much of what we have noticed in our perusal of chemical quantum mechanics and also points out how valence theories fulfil another important criterion of a good model which we have repeatedly come across, namely that within their terms of reference models should be able to predict.

We have now discussed three of the classes of model that carry the main burden of the organic chemist's thinking, the functional group, effects and quantum mechanics. The fourth, reacting systems, will be taken up in the next chapter, but we look first, though in brief, at modelling in the conceptual thinking of physical and of inorganic chemistry.

3.3 **Modelling in physical chemistry**
3.3.1 *The scope of physical chemical methods*

Chemistry regarded from a physical standpoint with the fundamental properties of atoms and molecules treated as quantitatively as possible, responds to modelling in a very basic manner. Essentially each system under 'study is well-defined (in contrast to some of the biological systems in chapter 6) and its behaviour under the influence of a perturbation may be described in terms of a model of the system. This model is usually designed to facilitate the construction of a useful mathematical expression or a correlation of the parameters involved. The situation is analogous to the quantum mechanical models we have just discussed: we are dealing with class one models. Indeed quantum mechanics may play an essential part in the model because all properties at the atomic and molecular level can be discussed in terms of quantum mechanics. For bulk systems, of course, classical mechanics offers a valid basis.

Physical chemistry plays an important part in building up and supporting the conceptual model of chemists in other branches of chemistry and on a larger scale is also at the very roots of chemical technology (chapter 7). It is the physical chemist who is asked about bond angles and bond lengths, heats of formation and other molecular quantities used by the theoretician; he is consulted by experimental chemists and technologists on bulk properties such as heats of reaction and solvent properties. Thus at every turn physical chemistry is part of the basis of chemical practice and of the chemical conceptual model.

As we mentioned just now, the modelling procedures discussed in the previous section can be applied to physical chemical problems. However since physical chemistry provides the experimental data for theoreticians and others to work from, it is clear that great importance must be attached to the physical meanings of parameters introduced into physical chemical models. We have already seen an example of the significance of parametrisation in the discussion of the gas laws in chapter one. This is a typical physical case in that it is built upon empirical correlation of observation and subsequently improved with physically interpretable parameters to agree better with experimental data. In this light let us first trace the history of another well-developed branch of classical physical chemistry, the theory of electrolytes, and subsequently note the many areas of modelling work which have a physical chemical background and are discussed in other contexts in other parts of this book.

3.3.2 *The theory of electrolytes*

The theory of electrolytes as developed since 1883 principally by Arrhenius, Debye, Hückel, Onsager and Bjerrum began from the experimentally observed fact that the conductance of a solution of an electrolyte varies with the concentration of the electrolyte. Between 1883 and 1887 Arrhenius (ref. 25a) drew up a model which based conductance upon an equilibrium between ions and undissociated molecules from which the conductance at concentration c, λ_c, could be expressed as a function of the degree of dissociation of the electrolyte α (equations 3.11).

$$AB \rightleftharpoons A^+ + B^-$$

$$\alpha = \frac{\lambda_c}{\lambda_\infty}$$

(3.11)

where λ_∞ is the conductance at infinite dilution. The degree of dissociation could also be determined independently from the colligative properties of the solution (osmotic pressure, melting point depression etc.). Good agreement was found for electrolytes which are composed of large ions with a low charge (weak electrolytes, commonly containing organic ions) but the agreement for strong electrolytes (ions with a large electric field around them) was very poor.

At the turn of the century, then, the demons in the theory were associated with the large electric field of highly charged ions and modifications to account for this were required. These improvements were made by Debye and Hückel in 1923 (ref. 25*b*) and by Onsager in 1928 (ref. 25*c*). They were influenced by the early discoveries of X-ray crystallography which showed that strong electrolytes are ionic in the solid state and are completely dissociated. They therefore suggested that ions of strong electrolytes are also completely dissociated in solution and went on to propose the following model. The ion X^+ in a solvent of high dielectric constant is surrounded by an atmosphere of other ions drawn to it by its electric field and consisting largely of ions of the opposite charge Y^- (figure 3.7). The motion of the ions, which

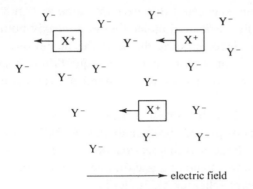

Figure 3.7. The ion atmosphere of cation X^+ in an electric field.

gives rise to the conductivity, can then be seen to slow down by two effects of the ion atmosphere. The ion atmosphere tends to drag X^+ back with it when an external electric field is applied and, in order to move towards the negative electrode, the X^+ ion must break free of its ion atmosphere into a new one. This model assumes that the solvent is a uniform continuum.

This is the conceptual basis for the development of a mathematical model of the phenomena. The effect of each perturbation mentioned above on the Coulomb energy of an ion was related within the framework of the ideas outlined above and an equation of the form 3.12 was derived.

$$\lambda_c = \lambda_\infty - (A + \lambda_\infty B)\sqrt{c}. \tag{3.12}$$

A and B are composite constants of such intuitively important quantities as solvent dielectric constant and viscosity, charge on the ions etc. They derive naturally from the equations that the model uses to estimate the Coulomb energy of the system and are not disposable parameters requiring an empirical assignment of their numerical value (compare the constants in the models discussed in section 3.2.2 for evaluation of strain in molecules). Notice the parallel development of this model with the development of the conceptual model in general – progress in the understanding of the structure of solids was needed in order to gain an insight into the interactions which might be possible in solution.

The model expressed by equation 3.12 was successful for uni-univalent electrolytes but more highly charged ions again deviated from its predictions. Arrhenius, the originator of the theory, had the last laugh when Bjerrum suggested that perhaps some ions may associate to form not a fully associated molecule but a weakly bound ion pair (refs. 25*d*, 26). By incorporating this idea the model once again becomes a close relative of the original single equilibrium system of Arrhenius. Ion pairing has since been well authenticated in organic and inorganic chemistry.

The reader interested in the physical aspect of models will find many further examples dispersed throughout this book. In particular, physical models play a large part in spectroscopy (chapter 5), kinetics and reaction mechanisms (chapters 4 and 6), and chemical technology and the design of a plant (chapter 7). In addition relevant comments will be found in chapter 6 where biological phenomena of physical interest are discussed.

3.4 The conceptual model in inorganic chemistry

3.4.1 *The relationship between inorganic and organic chemistry*

We shall see in chapter 6 which discusses modelling in systems of biological interest the development of inorganic chemistry into a field which was previously the concern of biochemists and a very few organic chemists. This 'blurring' of interdisciplinary borders could not have occurred if organic and inorganic chemists spoke entirely different languages. It is rather as if both traditional disciplines were encouraged to tune their ears to understand each other's dialect. Thus there is no real difference between the conceptual models of organic and inorganic chemistry. A continuum exists through which the emphasis on particular aspects varies.

Inorganic applications of MO theory, for example, may have close parallels with organic cases. One can use the quantum mechanical MO method to describe the bonding of the well-known sandwich compounds such as ferrocene (ref. 27). Ferrocene contains an organic component (cyclopentadienyl anion) which can be deduced from an 'organic' MO method such as the Hückel method (ref. 18*e*) (figure 3.8). This example demonstrates the strong bridge between organic and inorganic chemistry that is found in organometallic chemistry and the importance of the field stretches far beyond the walls of the research laboratory. Important industrial processes such as the Ziegler–Natta olefin polymerisation processes are based on organometallic catalysis.

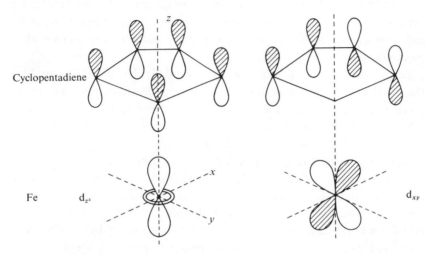

Figure 3.8. A MO treatment of ferrocene.

3.4.2 *Systematisation in inorganic chemistry*

In some respects the inorganic dialect contains a greater range of expression than the organic, which is possibly more rich in nuance and subtlety. Inorganic chemistry, of course, is concerned with all the elements, whereas organic chemistry deals mainly with carbon and the elements that commonly bond to it. One of the first models of chemistry upon which much of the present conceptual model has been built was introduced in the field of inorganic chemistry. This was the periodic classification of the elements (Mendeleef, 1890s) and it could be said to form the basis of the inorganic chemist's conceptual model. Today this is almost taken for granted since the periodic classification has been shown to be the natural outcome of quantum mechanics.

3.4.3   *Class one models in inorganic chemistry*

Inorganic chemistry possesses almost as wide a variety of structural arrangements about a single atom as it does elements. The wider variety of orbitals available for bonding (d and even f orbitals as well as the s and p orbitals common in organic chemistry) is responsible for this abundance and, although the structures and properties can be described by MO methods, albeit with some difficulty, inorganic chemists have been fond of discussing structures in terms of the simpler ligand field theory (ref. 18*f*). This theory has, however, been replaced by more modern models and Jørgensen (ref. 28) has described seven distinct models for the octahedral MX_6 molecule. He evaluates each model thoroughly by comparing predictions with experimental data such as the spectrochemical series which we shall meet shortly.

In its simplest forms known as crystal field theory, ligand field theory (LFT) is semi-quantitative, neither rock-solid quantum mechanics (class one), nor solely effects which are difficult to quantify (class three). Briefly LFT describes the effect of bonding of different ligands to a transition metal ion on the electronic structure of the ion, in other words on the energy levels of its d orbitals (ref. 18*f*).

In an inorganic complex the ligands are found at defined sites around the metal ion and the metal d orbitals point radially in their general direction. These orbitals are of equal energy in a hypothetical 'free' metal ion but, when they interact with a ligand orbital full of electrons they are repelled by the electrostatic interaction and are thereby raised in energy.

In an octahedral complex, for example, the ligands point directly along the axes of the $d_{x^2-y^2}$ and the d_{z^2} and accordingly these two orbitals are destabilised with respect to the other three d orbitals (d_{xy}, d_{yz}, d_{xz}) which are stabilised (figure 3.9). Different geometries of complex produce different d orbital splittings since the ligands are oriented towards different orbitals. The magnitude of the separation between two groups of d orbitals, or the ligand field splitting as it is called, varies from metal ion to metal ion, but the splitting power of ligands follows a fairly rigid sequence which is known as the spectrochemical series.

One contribution of LFT to the conceptual model is to provide a visualisable understanding of much metal complex chemistry in terms of the magnitude of the ligand field applied. One could perhaps

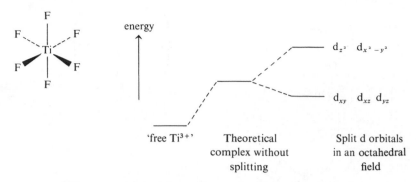

Figure 3.9. Ligand field-splitting in an octahedral complex.

compare the modelling aspects of LFT with the Hammett correlation of substituent effects (section 3.2.2). In each case relevant experimental data are known but it is difficult to account for the results in terms of more fundamental components than the observed parameters (σ and ρ, and the ligand field splitting). At a more fundamental level LFT has proved to be very useful as a basis for quantum mechanical calculations of metal complexes.

One reason for the development of the HSAB model (section 3.2.3) was to provide a more overt unity between inorganic and organic chemistry. The case of carboxypeptidase discussed in chapter 6 is another unifying example. Here we notice that only the hard metal ions Co^{2+}, Zn^{2+}, Ni^{2+}, and Mn^{2+} can attack the essentially hard carbonyl group of the substrate; the soft metal ions Cd^{2+}, Pb^{2+}, Hg^{2+}, and Cu^{2+} all fail.

3.4.4 *Class three models in inorganic chemistry*

Most commonly, inorganic chemists specialise in the chemistry of one group of elements or even in the chemistry of one element alone. Since elements and ligands clearly belong to class two of the conceptual model their interactions can be mediated by effects, members of class three. The inorganic effects are of the same type as those discussed in section 3.2 with the addition of ligand field effects. The direction of electronic effects is, as before, controlled by the electronegativities of the elements, and much of the chemistry can be discussed in quantum mechanical terms. To conclude this chapter, three examples of the operation of effects in inorganic chemistry will be discussed.

Acetylacetone forms very strongly bound chelates with many

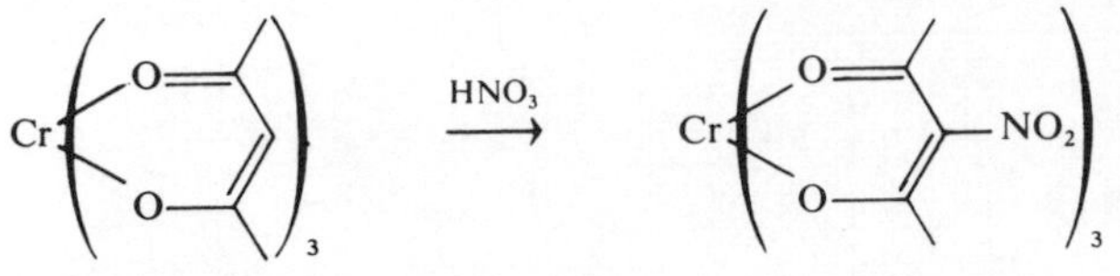

(3.10.1)

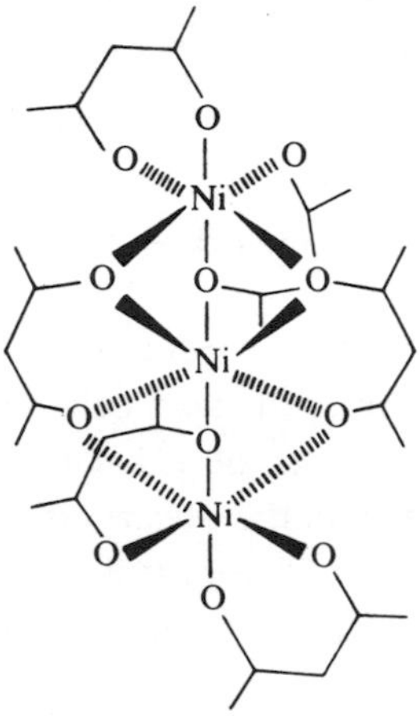

multiple bonds omitted for clarity

(3.10.2)

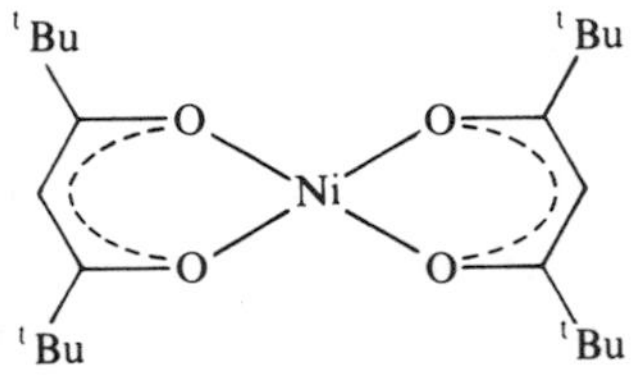

(3.10.3)

Figure 3.10. Examples of effects in inorganic chemistry.

transition metal ions including chromium (III). The violet chromium (III) complex reacts with nitric acid to give a nitro derivative without decomposition or oxidation of the chelate (3.10.1). This is because the electrons donated from the ligand into the 3d orbitals of chromium form a six-membered ring with pseudo-aromatic character. Thus the methine C–H bond can be substituted like benzene without decomposition of the chelate (ref. 29).

The foregoing example shows the stabilising effect that a metal can have on a ligand. Equally striking effects can occur from the opposite direction. A property of metal ions that is very sensitive to the ligand is the redox potential and a typical case is the ferrous–ferric couple. The table shows some values of the redox potential for iron compared with a representative group of ligands (table 3.2) (ref. 30).

Some of the reasons for this trend are not hard to perceive from the conceptual model, in this case without the need for the subtlety of ligand field effects. The more highly charged ferric ion would be expected to be stabilised by the more anionic ligands (CN^-, $EDTA^{4-}$)

Table 3.2. *Ligand effects on the redox potential of iron* $(Fe^{2+} \rightarrow Fe^{3+} + e^-)$

Ligand	Redox potential V
(o-phen)$_3$	-1.06
(H$_2$O)$_6$	-0.77
(CN$^-$)$_6$	-0.36
EDTA^{4-}	$+0.12$

and conversely the interaction of the π-orbitals of phenanthroline with the ferrous ion reduces the electron density around the metal and thus has a stabilising influence on the ferrous state.

Steric effects, common in organic chemistry, can also be observed in the structure of inorganic complexes. Nickel(II) complexes, for example, have several geometries open to them including octahedral and square planar. Which structure is found depends upon the ligand. The green trisacetonylacetonate complex, which in fact exists as a trimer (3.10.2) is octahedral whilst the analogous t-butyl substituted derivative is square planar and typically red (3.10.3). Obviously the extreme bulk of the t-butyl groups prevents the formation of the octahedral complex which is more stable for the ligand field strength of acetylacetone (ref. 31).

From the above examples it is apparent that our assertion that the conceptual model of chemistry as we have analysed it applies as much to inorganic compounds as it does to organic compounds. As our exploration of chemical reactions develops in the next chapter, further close parallels will be found, parallels both of chemical behaviour and of the models used to interpret and rationalise it. In fact the influence of the conceptual model will be seen as we progress to extend beyond the traditional boundaries of chemistry into industry and biology.

MODELS OF REACTING SYSTEMS

4.1 Introduction

Chemistry is about the structure of molecules and how one molecule may be converted into another. Accordingly, models of chemical reactions must describe what happens when two molecules come together and react. As soon as we enlarge the system under consideration to include several species of molecule we are creating a new environment for each of them. This is a change in the external environment which may not only affect the rate of a chemical reaction, but may also change its course. We shall find that a blend of iconic and symbolic models has been constructed to describe and predict the behaviour of reacting systems.

4.2 Describing reactions through models

So far we have said nothing about the rate at which a chemical reaction occurs or indeed how it occurs. Both of these facets can be illuminated with the aid of models and clearly the two are interlinked. Let us begin with a model that we first met in chapter 1, the kinetic theory of gases. The discussion of the early theories of reactions that follows is chemically unsophisticated; readers will probably be familiar with most of the material.

4.2.1 Collision theory

The kinetic theory of gases postulates that gases are comprised of molecules which are hard spheres in rapid, random motion and which collide with each other at intervals. As a result of these collisions, they undergo only physical changes, such as a change in velocity. We can readily extend this system to include two types of molecules within the gas which on collision, one type with the other, interact chemically as well as physically to form a new species. Intuitively the rate of such a process must be related to the frequency of the collisions. Since this frequency can be estimated from kinetic theory, we now have a basis for a theoretical model that can predict and explain the rates of chemical reactions in the gas phase.

Notice the restrictions that are placed upon this model at the outset. By limiting ourselves to reactions in the gas phase we eliminate the possibility of solvent effects and can concentrate solely upon the reactions between the two molecules. Of course we must, as always, take care not to ignore potentially significant factors. For example, it could be that a given reaction takes place only on the walls of the vessel because of some catalytic interaction. Likewise we could be mistaken in believing that there are only two molecular species present; a minute trace of water might be present and essential for reaction. Although such factors are common in gas phase reactions, it is better in the first instance to construct a simple model based upon an idealised system and to bear these potential sources of difficulty in mind when running and testing it.

The collision theory of reactions was the first quantitative theory of reaction rates to be propounded (*c.* 1890, ref. 1*a*). There are several ways of describing the collisions that take place between reactant gas molecules, depending upon whether such factors as electric charges upon either or both reactants affect the probability of a collision, but in all cases the reaction rate is considered to be a function of the frequency of collisions. However, it quickly became evident that this simple picture is inadequate. Each molecule of the hard sphere type collides about 10^{28} times per second at room temperature and pressure. This very high frequency implies that such gas phase reactions should be virtually instantaneous but this is not the case. Experiments showed that the rate of reaction often increased with temperature and this led Arrhenius in 1889 to suggest that a certain energy, which he called the activation energy, E_a, must be attained by molecules before they can react. This energy can be acquired by the transfer of kinetic energy during collisions and clearly, the faster molecules move (i.e. the hotter they are), the greater the chance that the activation energy will be exceeded and that reaction will occur (ref. 1*b*).

Arrhenius arrived at these ideas by pointing out a now common analogy that we met in the previous chapter's discussion of the Hammett Equation. He postulated that, since the equilibrium constant can be written as the ratio of the rates of the forward and back reactions, then it is reasonable that equations of the same mathematical form will describe the variation of both rate and equilibrium constants with temperature. Van't Hoff had already provided equation 4.1 as a relationship between the internal energy change of a reaction (ΔU), the equilibrium constant (K), and the temperature (T):

$$\frac{\mathrm{d}\ln K}{\mathrm{d}T} = \frac{\Delta U}{RT^2}. \tag{4.1}$$

Arrehenius simply wrote down the directly analogous equation (4.2) in which k is the rate constant:

$$\frac{\mathrm{d}\ln k}{\mathrm{d}T} = \frac{E_a}{RT^2}. \tag{4.2}$$

If E_a is independent of temperature, equation 4.2 can be integrated and rewritten as equation 4.3, the famous Arrhenius equation:

$$k = A\exp\left(\frac{-E_a}{RT}\right). \tag{4.3}$$

A is the constant of integration and is known as the pre-exponential factor. The exponential term is a probability function, related to the Boltzmann distribution function, which indicates the fraction of the molecules that have managed to attain the activation energy or greater.

So expressed, collision theory accounts remarkably well for the rates of such simple reactions of small molecules as

$$H_2 + I_2 \rightarrow 2HI$$
$$2NO + O_2 \rightarrow 2NO_2$$

and the Arrhenius equation provides a reasonably good description of the temperature dependence of the rate of these reactions.

The predictions of collision theory assessed. Recall now that simple kinetic theory takes as its model for molecular shape a simple hard sphere. This implies that all the axes of the molecule are identical and consequently that molecular orientation at the moment of collision will be of no significance. In fact, the sphere is not a satisfactory model of any molecule other than the smallest. Moreover, it turns out that if a reaction is to occur then molecules must collide not only with sufficient energy, but also in a particular relative orientation. It follows that simple kinetic collision theory usually predicts reaction rates that are far too high. To meet this difficulty, an empirical fudge factor p, known as the steric factor, was introduced into the rate equation to give equation 4.4

$$k = pA \exp\left(\frac{-E_a}{RT}\right). \tag{4.4}$$

The values found for p by comparing the experimental rate constants with the calculated ones are eloquent.

Reaction	p
$H_2 + I_2 \rightarrow 2HI$	0.33
isobutene + HCl → t-butyl chloride	3×10^{-6}

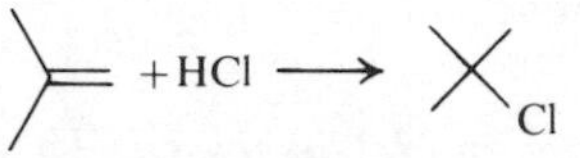

The reaction of hydrogen and iodine to give hydrogen iodide may be adequately described in terms of a broadside on collision between molecules as illustrated in figure 4.1 (a modern reassessment of the

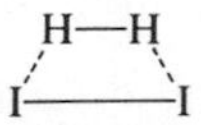

Figure 4.1

situation is discussed in section 4.7). This approximates fairly well to the hard sphere model of molecules. However for a molecule like isobutene, which is essentially trigonal, such a simple picture is clearly not appropriate. Naturally a more realistic model was sought, and one that did not rest upon unexplained empirical correction factors. Thus in the 1930s the theoretical model of reactions known as transition state theory was developed (ref. 1c) and it has provided the basis for most discussions of reaction rates ever since.

4.2.2 Transition state theory

Most chemical reactions involve the breaking and making of chemical bonds. Transition state theory, or activated complex theory as it is also known, singles out the bond breaking step for special consideration. As in collision theory, the reactant molecules collide, but rather than forming the products as a direct result of the collision, the reactants form a species known as the transition state or activated complex which is in equilibrium with the reactants and which can then proceed to form the products.

$$A+B \rightleftharpoons (AB^*) \rightarrow \text{products.}$$

The reaction rate is considered to be proportional to the concentration of the activated complex.

The transition state is like a normal molecule except that one bond, the bond that is to break, is in a fully excited vibrational state and this vibration is on the point of converting itself into a translation; conversion of vibrational energy into translational energy causes the bond to break and the resulting molecular fragments, the products, fly apart. Reaction has then occurred. It is obvious that the pathways along which a process like this can occur are very strongly dependent upon the structure of the reactants.

The predictions of the transition state model assessed. In order to calculate the rate of a reaction, transition state theory uses either statistical mechanics or quantum mechanics to calculate the properties of the transition state and reactants. The predictions are almost always better than those given by the hard sphere collision model as the following values for the pre-exponential factors of some bimolecular gas phase reactions show (ref. 1*b*).

Observed and calculated pre-exponential factors

Reaction	Observed	$A \, cm^3 \, mol^{-1} \, s^{-1}$	
		Collision theory	Transition state theory
$NO_2+CO \rightarrow NO+CO_2$	13.1	13.6	12.8
$NO+O_3 \rightarrow NO_2+O_2$	11.9	13.7	11.6
$2NO_2 \rightarrow O_2+2NO$	12.3	13.6	12.7
$2ClO \rightarrow Cl_2+O_2$	10.8	13.4	10.0
$NO+Cl_2 \rightarrow NOCl+Cl$	12.6	14.0	12.1
$2NOCl \rightarrow 2NO+Cl_2$	13.0	13.8	11.6

Notice that the hard sphere collision theory, which treats every reactant molecule as the same except for its mass, produces very similar results irrespective of the reaction and that in every case the result is an overestimate. On the other hand, transition state theory, which takes account of the geometry of the reactants, produces answers in good agreement with the experimentally determined values. The only calculation seriously in error is the last one listed and this can probably be ascribed to an error in the model of the transition state that was used, rather than to an invalid theoretical method.

Transition state theory, moreover, provides a satisfactory theoretical interpretation of the pre-exponential factor in the Arrhenius equation without recourse to empirical fudge factors in the following way. We have seen that the activated complex is considered to be in equilibrium with the reactants. Transition state theory relates the rate constant for a bimolecular reaction (k) to the equilibrium constant for the formation of the activated complex ($K^{\ddagger}$) thus:

$$k = \frac{k_{B} T}{h} K^{\ddagger}, \tag{4.5}$$

where k_{B} is the Boltzmann constant, h is Planck's constant and T is the absolute temperature. By applying the standard thermodynamic relationships:

$$\Delta G^{\ddagger} = - RT \ln K^{\ddagger},$$

and $\quad \Delta G^{\ddagger} = \Delta H^{\ddagger} - T \Delta S^{\ddagger},$

the expression for the rate constant may be written:

$$k = \frac{k_{B} T}{h} \exp \left(\frac{\Delta S^{\ddagger}}{R} \right) \exp \left(-\frac{\Delta H^{\ddagger}}{RT} \right) \tag{4.6}$$

where $\Delta G^{\ddagger}$, $\Delta H^{\ddagger}$ and $\Delta S^{\ddagger}$ are respectively the free energy, enthalpy and entropy of activation. Compare this equation (4.6) with the collision theory equation 4.4.

$$k = pA \exp \left(\frac{-E_{a}}{RT} \right). \tag{4.4}$$

Clearly $\Delta H^{\ddagger}$ corresponds to the Arrhenius activation energy, E_{a}, and the pre-exponential factor, pA, is related to the entropy of activation, $\Delta S^{\ddagger}$. Entropy is connected with the state of order in a system – an increase in order results in a decrease in entropy. Consequently, the magnitude of $\Delta S^{\ddagger}$ tells us a good deal about the nature of the transition state and in particular it allows low steric factors to be explained. From a comparison of equations 4.4 and 4.6 a low steric factor must be related to a small entropy of activation and therefore an increase in order must an increase in order must occur on forming the activated complex. Such an argument could be applied to the addition of hydrogen chloride to isobutene where intuition told us that something more tightly structured than a broadside collision is required for reaction.

Models derived from transition state theory. In addition to the quantitative descriptions that we have just discussed, a number of more pictorial models of transition states are in common use, especially when large molecules are concerned. We shall have more to say about detailed models of transition states shortly but a few examples now will help to clarify what follows. At the most trivial, the transition state can be understood as a 'half-way-house' between reactants and products where the appropriate bonds are half made and half broken, for example as indicated for the exchange of protium and deuterium:

$$D+H–H \rightarrow D\ldots.H\ldots.H \rightarrow D–H+H.$$
transition
state

A much more useful representation is the reaction profile (figure 4.2). This plots the changes in energy during a reaction against the extent to which the reaction has occurred (reaction coordinate); the plots are smooth curves in which the transition state lies at a maximum. Reaction profiles are valuable pictorial models and they can be calculated easily for small molecules with only one covalent bond by use of quantum mechanics. Further, it is possible to plot the potential energy of a reacting system such as that of deuterium with hydrogen in terms of defined geometrical coordinates. When dealing with a triatomic reaction, one atom is considered to be stationary and the other two move relative to it. In this way, a map of the reaction path and its surroundings is obtained. The map represents the potential energy of the system as contour lines (figure 4.2) and the topography thus shown is a potential energy surface for the reaction which allows us to follow the reaction's course in detail. Surfaces have been calculated for a number of reactions including hydrogen and deuterium exchange (refs. 1*d*, 2). In this case, the reaction follows the potential valleys leading up to a saddle or col (the transition state) from which it rolls downhill to the products.

If we wish to proceed to more complex molecules, we run up against the limitations of quantum mechanics once again: it is not possible to calculate the potential energy surfaces for complex reac-

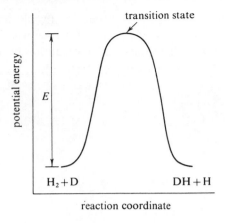

transition state

potential energy

E

H₂ + D   DH + H

reaction coordinate

Reaction profile

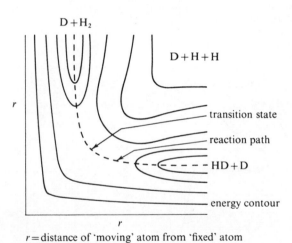

D + H₂

D + H + H

transition state

reaction path

HD + D

energy contour

r

r = distance of 'moving' atom from 'fixed' atom

Potential energy surface

Figure 4.2

tions without vast expense of computer time or the use of semi-empirical methods (section 4.6). For the thermal isomerisation of cyclopropane however, a unimolecular reaction with twenty-one dimensions, the calculation of the potential energy surface has been brilliantly achieved (ref. 3) but under current limitations such efforts cannot become common. Nevertheless, the symbolic and pictorial models that transition state theory has provided are valuable aids to the understanding of different types of reactions and can be manipulated as qualitative components of the conceptual model. Thus the reaction profile for a strongly exothermic reaction must be like (4.3.1) and a reaction that passes through a metastable intermediate must have a reaction profile like (4.3.2). Such profiles, like the entropy of activation, imply

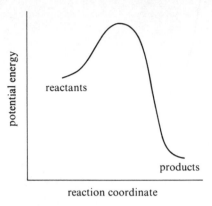

(4.3.1) An exothermic reaction

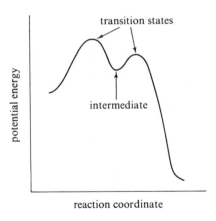

(4.3.2) A reaction involving an intermediate

Figure 4.3

relationships between the structures of the reactants, intermediates, transition states, products, and enable us to postulate pathways for reactions without the need to calculate them precisely. Reaction pathways that include descriptions of the motion of atoms and electrons are commonly called reaction mechanisms.

Transition state theory makes use of a number of assumptions to simplify the calculation of rate constants. One such simplification is to assume that the vibrational energy of the bond to be made or broken is independent of the vibrations of the rest of the molecule. Experimental evidence shows that this is only approximately true: for example the magnitude of a kinetic isotope effect is often less than the application of simple transition state theory would predict (ref. 4*a*). The bond that is made or broken at the transition state is thus another example of a quasi-independent system.

Transition state theory has many of the virtues of a good theoretical model in that it provides a flexible, intuitively attractive base to work on either quantitatively or qualitatively. It is also a fertile soil in which to germinate more specialised theories of reactions as we shall see later in this chapter.

4.2.3 *Reaction mechanisms*

In describing a transition state for a reaction we are beginning to talk not only about the rate of the reaction but also about the path it follows. Any description of a reaction path can be called a mechanism, but the particular description chosen depends very much upon the purpose of the study in hand. We shall present examples in which the detailed energetic and geometric changes in a molecule have been calculated using quantum mechanics and on the other hand we shall demonstrate the simple mental exercise of applying empirical generalisations to try to predict the course of a reaction. The mechanism is a model of the behaviour of the reacting system that can be expressed in any of the symbols and concepts that we have described.

Apart from giving the satisfaction of an understanding in familiar terms of how a reaction proceeds a good mechanism like any scientific theory of value must be open to test by experiment. Most importantly it must also supply the chemist with a predictive tool that will allow him to make a better guess at what will happen in a reaction than he could make simply by considering effects and functional groups, a procedure that may neglect the influence of the outer environment. This is particularly true when subtleties of reactions such as the stereochemical course are important.

In this chapter a range of mechanisms of general applicability in organic, inorganic and biological chemistry will be discussed. To emphasise the generality of modelling we have chosen to describe models as they are applied at each stage in a reaction. Firstly we look at the ways in which molecules are described by structural formulae and at the implications of these formulae. With an understanding of this we then turn to examine influences exerted by the outer environment such as solvent effects, salt effects, and catalysis. This will in turn lead to a discussion of models that attempt to describe the properties of the transition state and then lastly we take stock in a general way of the common situation when there is scant experimental data to aid the proposal of a mechanistic working hypothesis.

Mechanisms could equally well be classified and discussed according to the reaction type that is in question. This approach, which will be taken in chapter 5 can lead, however, to the undesirable tendency to concentrate upon either organic or inorganic reactions and in this chapter we wish to take a broader view.

One exception to this stricture is the use of labels such as S_N1 and S_N2 which designate respectively unimolecular and bimolecular substitution reactions. These and related reactions are common to both inorganic and organic substrates and they supply us with many pertinent examples in this chapter. The terms S_N1 and S_N2 are effectively short-hand expressions for classes of mechanisms; under each can be grouped a number of characteristic experimental observations. These relate to such phenomena as salt effects, substituent effects, kinetics, and stereochemistry, all of which are aspects of the reaction mechanism. Labels like these are valuable aids to the chemist in organising his material and an aid to the easy assimilation of mechanisms into a conceptual model.

4.3 Structural formulae

Structural formulae are not only indispensable as an aid to communication but also are important as a template from which the chemist can develop his own personal mental picture of a reaction. Words alone cannot describe the relationship between atoms in a readily comprehensible way, neither can equations encompass enough information for most chemists. For these reasons, convenience and custom dictate that communication between chemists be conducted mostly with the help of iconic models, namely drawings in two dimensions. The two dimensional drawing is still the most practicable and common representation despite the increasing use of computer-generated stereoscopic drawings in fields such as X-ray crystallography, where three dimensional relationships are of paramount importance. The stereoscopic technique is especially revealing in the study of enzymes and the mechanisms of enzymic catalysis (chapter 6).

In addition to the conventional representations of molecules that we have already used there are a number of other forms of notation for molecules. Many of these have been designed for use in computer systems. The Wiswesser Line Notation (WLN), for example, can be used for compiling a computer based index of compounds. The notation is in full accord with our description of the conceptual model because it bases its description upon elements and functional groups. The Institute for Scientific Information advertises its WLN based structure index, 'For each compound a separate entry is created for every significant sub-structure it contains.' Figure 4.4 gives an example of a conventional structure (4.4.1) and its WLN counterpart (4.4.2).

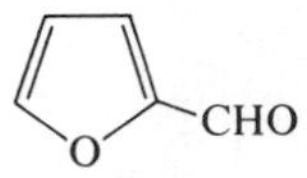

	T50J BVH
(4.4.1)	(4.4.2)

Figure 4.4

There is a close relationship between this approach and the application of topological theory to the structure and properties of molecules. Here chemical structures are encoded in matrices which map the interconnections of various atoms as topological indices that are particularly convenient for computer processing. Rouvray (ref. 5) gives an enthusiastic account of much of the leading work in this field although the reader may also find that inadequate attention has been given to the dangers of over-simplification. It is claimed, for example, that topological indices may have application to cancer research. Indeed they may, but under what circumstances might this be so? What assumptions would have to be made? It would be an interesting problem for the reader of this book to assess the relevance of the topological approach to cancer research on the basis of the current state of the art described by Rouvray.

Other forms of notation, particularly developed for computers, include several which have been designed to aid the devising of synthetic routes to complex organic molecules (see chapter 5). Firstly they allow the computer to generate synthetic steps by means of an exhaustive examination of all the possible precursor molecules that can be generated systematically by breaking one bond of the target molecule at a time. Secondly, they provide a means for the translation of the operations performed by the computer into classical structures thus allowing the chemist, who is programmed in terms of classical structures, easy communication with the computer (ref. 6).

4.3.1 *Reflections of quantum mechanics in structural formulae*

The development of structural representations has closely followed the advance of theory and the conceptual understanding of chemistry. As an example, let us compare the drawings which have at various times been proposed as representations of the structure and hence the chemistry of benzene (ref. 7) (4.5.1–5). The results of decades of research now support only the forms (4.5.1, 2 and 4) as acceptable representations of benzene. The others are now known to represent other compounds in their own right.

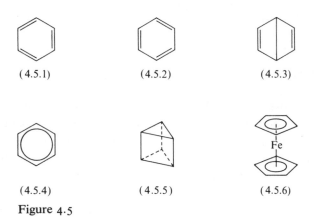

<table>
<tr><td>(4.5.1)</td><td>(4.5.2)</td><td>(4.5.3)</td></tr>
</table>

<table>
<tr><td>(4.5.4)</td><td>(4.5.5)</td><td>(4.5.6)</td></tr>
</table>

Figure 4.5

The Kekulé forms (4.5.1 and 2) and the MO form (4.5.4) emphasise that the two major theoretical bases of the conceptual model, MO and VB theories, require slightly different pictorial representations and as a consequence different interpretations. Classical structural formulae are easily reconciled with the VB two electron bond model of molecular structure: each line in a classical structure represents one two-electron bond. On the other hand, in order to represent a molecule like benzene in which MOs clearly govern the chemistry, it becomes necessary to use diagrams which give molecular orbitals due weight such as (4.5.4) or to resort to the use of a number of VB structures of similar energy which considered together represent the chemistry of the molecule. Such forms are illustrated by the Kekulé structures for benzene (4.5.1 and 2) and are known as resonance or canonical forms. A resonance description however, becomes more difficult and more cumbersome when larger molecules are in question or when more complex types of bonding are found. The bonding in ferrocene (4.5.6) for example, is better illustrated by an MO picture as in benzene than by a set of canonical structures.

Molecules like ferrocene that lend themselves to an MO description are not readily represented in simple line formulae. With this in mind, Dewar has commented

> One of the problems which has arisen with the development of sophisticated MO treatments is the difficulty of relating such calculations to the traditional picture of molecules in terms of localised bonds, resonance energy etc. SCF (self consistent field) expressions for the total energy are far more complex and it is very much harder to see their relationship to traditional ideas. [Ref. 8.]

The problem of the boron hydrides. There are circumstances when a hybrid notation of MO and VB becomes valuable. In the notation introduced by Lipscomb (ref. 9) for the structures and chemistry of boranes, the concept of canonical forms of VB resonance theory is combined with the concept of a three-centre bond which is derived from MO theory (figure 4.6). Boron possesses only six bonding electrons and so it is not possible to write a conventional VB structure with each line representing a two-electron bond for a molecule like diborane B_2H_6. It was therefore suggested that a bond might reasonably be formed by overlap of three orbitals rather than two and that the resulting bonding orbital would contain two electrons. For diborane, we assume that boron uses some hybrid of s and p orbitals, a VB formalism, and that these hybrids overlap with the hydrogen 1s orbitals in two ways. Firstly boron–hydrogen two-centre bonds notated in the conventional way can form (4.6.1), and secondly boron–hydrogen–boron three-centre bonds or bridging bonds can provide the link between the two boron atoms (4.6.2). The notion of three-centre bonding is very much part of MO theory and it could be considered as similar to the overlap of adjacent carbon p orbitals in conjugated polyenes. Indeed the lowest π-MO of benzene could be regarded as a six-centre bond.

Higher boranes require a further elaboration of the bonding scheme to account for their empirical formulae. It is necessary to invoke boron–boron–boron three-centre bonds also and Lipscomb has distinguished two types. The open B–B–B bond is directly analogous to

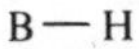

(4.6.1)

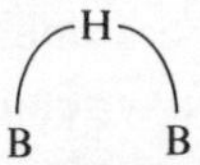

(4.6.2)

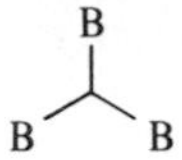

(4.6.4)

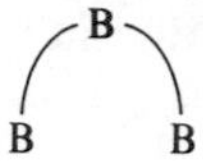

(4.6.3)

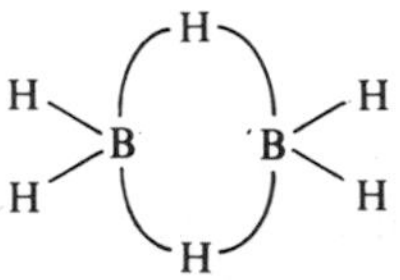

(4.6.5)

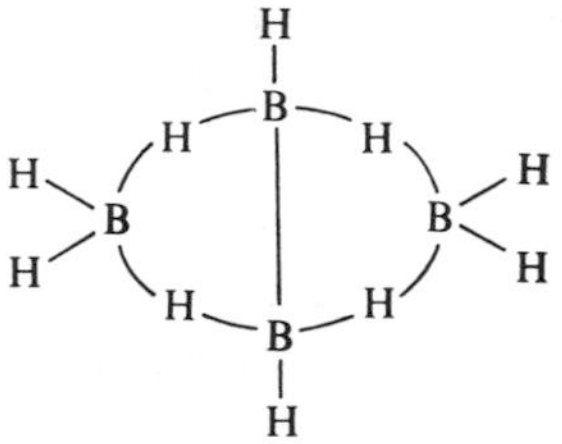

(4.6.6)

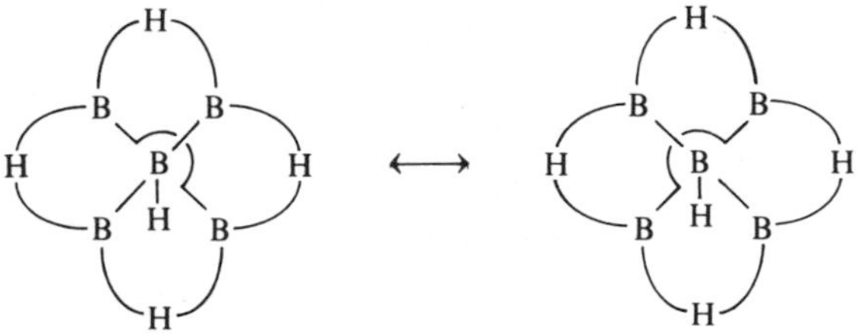

(4.6.7) (4.6.8)

Figure 4.6. Structure and bonding in some boron hydrides.

the B–H–B three-centre bond and is drawn as in (4.6.3). The second type, the closed B–B–B three-centre bond unites three boron hybrid orbitals to a common centre and is written as (4.6.4). Thus diborane can be represented as (4.6.5) and B_4H_{10} as (4.6.6). Still larger boranes introduce the possibility of more complex formulations. Thus pentaborane-9 could be written as either (4.6.7) or (4.6.8). On symmetry grounds we would expect both of these forms to have equal energy. They are thus analogous to VB resonance hybrids which we described for benzene. For this borane there are four more canonical forms and for decaborane-14 there are one hundred and eleven. When such complexity is reached VB structures become incomprehensible and the chemistry can be represented more usefully using MO theory assisted by the symmetry properties of the molecule.

Thus the range of pictorial descriptions for bonding in the boron hybrides parallels those available for organic systems. Carbenium ions also have six bonding electrons and are isoelectronic with boron. It is consequently no surprise to find three-centre bonds invoked in representations of 'non-classical' carbonium ions (ref. 10, see also chapter 3).

Some theoretical limitations of structural formulae. The use of structural formulae is successful because the Born–Oppenheimer approximation is valid for most chemical species (chapter 3). Provided that the nuclei of the atoms in a structure can be regarded as stationary with respect to the time scale of electronic movements, then conventional structural formulae are meaningful since the relative positions of the nuclei can be defined within the limits of vibrational motion. This holds for reactive intermediates and transition states as well as for ground state molecules. When the Born–Oppenheimer approximation does not apply it becomes difficult to provide a simple representation of the system. This is a problem in theoretical treatments of photochemical reactions, which involve the interaction of excited states as well as ground states. There are more reacting states possible than in ground state reactions and more structures are needed to define them (ref. 11). Consequently it is not easy to represent photochemical reactions by classical conventional structures which really illustrate the results of VB theory; most photochemical reactions are best described in MO terms.

4.3.2 Representations of functional groups in structural formulae

The risks of simplification by abbreviation. The notation for functional groups often uses abbreviations – a carboxyl group is abbreviated to CO_2H, alkyl groups to Me, Et, etc. and of course the general alkyl R. When these and similar abbreviations are used in a structure, the chemical nature of the compound that is represented is easily understood. There are, however, great dangers in over-abbreviating a structure. A part structure can easily omit a feature vital to the chemistry of the system under consideration and arbitrary abbreviations, such as the initial letters often used in biochemistry, have the disadvantage of giving no indication as to the chemical nature of the compound. An abbreviation should be sufficient to remind the reader of the chemical nature of the molecule in question even if he is unfamiliar with the precise structure that it represents.

We may note, as an example of the variety of abbreviation used to

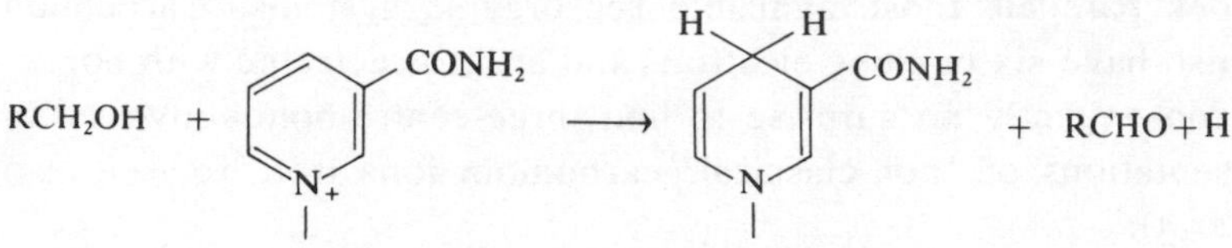

$$RCH_2OH \quad + \quad \underset{(4.7.1)}{\text{[pyridine ring with CONH}_2\text{]}} \quad \longrightarrow \quad \underset{(4.7.2)}{\text{[dihydropyridine ring with CONH}_2\text{]}} \quad + \quad RCHO + H^+$$

$$R.CH_2.OH \; + \; NAD^+ \qquad \longrightarrow \qquad NADH \; + \; R.CHO \; + \; H^+$$

Figure 4.7

represent the same reaction, that the reduction of the coenzyme nicotinamide adenine dinucleotide can be represented chemically as part structures by (4.7.1) and (4.7.2). Even these substantially abbreviated drawings clearly illustrate the nature of the chemical change that has occurred. Obviously in this example the full detail of the reaction cannot be shown: the stereochemistry of the reduction is omitted, for example.

Abbreviation of a structure is always accompanied by a loss of information and so the most suitable abbreviation depends upon the purpose in hand. Thus if in the above example, the substrate only is important, it is sufficient to represent the coenzyme by the initial letters NAD^+ or $NADH$.

Nowadays we take a pictorial structure of even the most complex molecule for granted and scarcely pause to think what is involved in the model that we are using. Chemists confronted with inconclusive experimental evidence have sometimes shown a deeper understanding than most of us who have spectroscopy to aid us. Robinson, a chemist who left his mark on the development of the conceptual model of organic chemistry, was very much in the former category when he wrote describing a proposal for the structure of morphine (ref. 12); 'This formula has been adopted especially since the formulae which it is suggested should replace it cannot without hesitation be accepted as superior expressions of the properties of the substance.'

Stereochemical problems. The problem of the representation of stereochemistry is very important. So far, we have considered representations of two dimensions only: thus planar benzene is very satisfactorily represented by (4.5.4) and a long chain fatty acid may be represented by $CH_3(CH_2)_7CO_2H$.

But consider the drawings in figure 4.8. The left hand formula

$CH_3 . CH(\overset{+}{N}H_3) . CO_2^-$

(4.8.1)

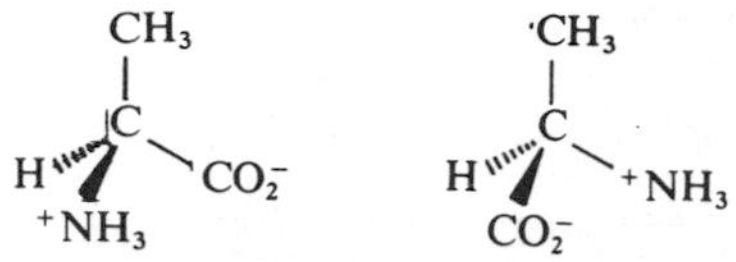

$Co(H_2NCH_2 . CH_2NH_2)_3^{2+}$

(4.8.2)

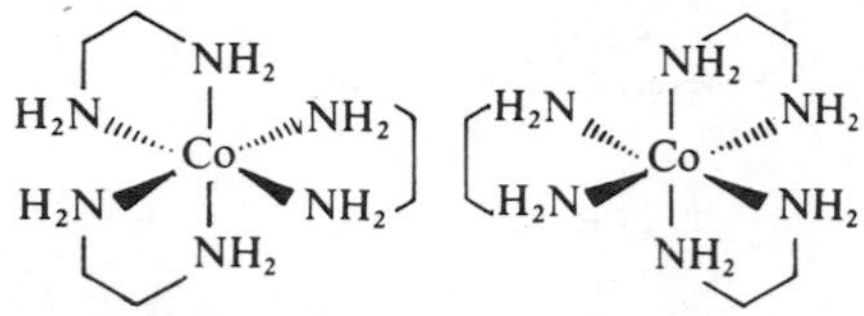

Figure 4.8

represents the same compound as the right hand ones, yet the right hand drawings tell us a very great deal more about each compound. The bare constitutions of (4.8.1) and (4.8.2) do not show that here we are dealing with chiral compounds. This means that it is necessary to have an efficient method of representing the three dimensional structure of molecules on paper and to do this, the chemist adopts conventional projection formulae (ref. 13). Examples of projections are shown in figures 4.8 (right) and 4.9. These drawings are essentially views of the molecules from different angles in which the usual two electron bond represented by a line is modified. Either the bond is visually strengthened as in the wedge shape which represents a bond projecting in front of the plane of the paper, or the bond is represented as a fainter dashed line thereby indicating a bond that projects behind the plane of the paper. The normal single line bonds are in the plane of the paper itself.

Drawings such as these are highly pictorial and easily interpreted with a little practice and with the aid of molecular models (see chapter 5). There are, however, other common projections used in organic chemistry that are more abstract and require the adherence to carefully defined conventions for use. In figure 4.9 we see one compound, butan-2,3-diol, represented in three different conventions, the Newman projection (4.9.1), the sawhorse projection (4.9.2) and the Fischer projection (4.9.3). Each of these projections, is a model that emphasises a particular stereochemical feature. The Newman pro-

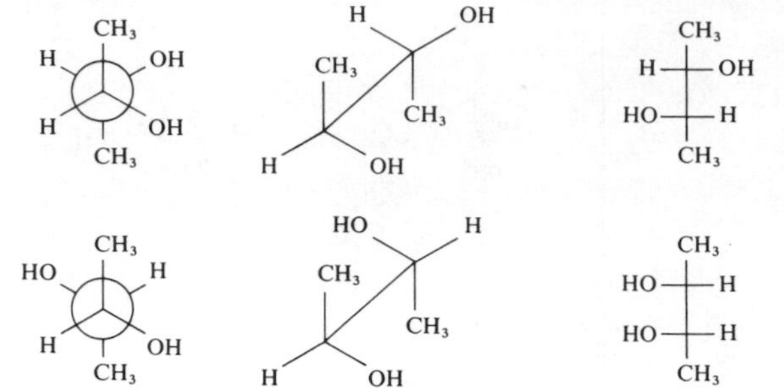

(4.9.1)
Newman projection

(4.9.2)
Sawhorse projection

(4.9.3)
Fischer projection

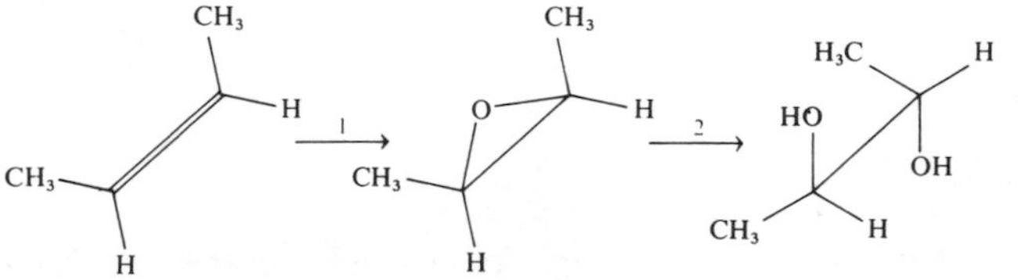

Reagents: 1. peracid, 2. aqueous base

(4.9.4)

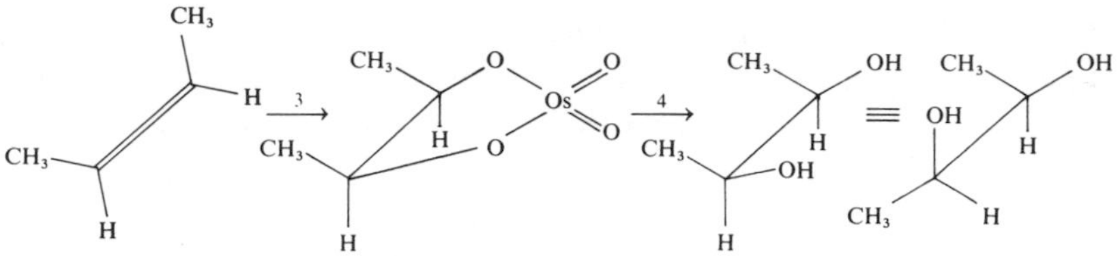

Reagents: 3. osmium tetroxide 4. dilute acid

(4.9.5)

Figure 4.9

jection is valuable for investigating the conformational properties in systems in which C–C bonds are free to rotate. The Fischer projection concentrates attention upon the configurational relationships between molecules (for example the ± and meso forms illustrated). The sawhorse projection, apart from being useful to aid transforming a Newman projection into a Fischer projection, allows the chemist to depict the stereochemical changes that occur at two adjacent carbon atoms during a reaction and hence to predict the stereochemical course of the reaction. The oblique view taken by the sawhorse projection makes it easy to see that conversion of cis-but-2-ene to butan-2,3-diol via the epoxide (4.9.4) and hydrolysis will give rise to the meso form.

In contrast, the same conversion using osmium tetroxide would afford the ± diol via the osmate ester (4.9.5).

Structural formulae are among the most important keys to the conceptual language of chemistry. To use them effectively a chemist must be able to draw the structures of the molecules that he is working with and to extract from the drawing the qualitative chemistry to be expected from the system. Stereochemistry may well be important. Furthermore a reaction involves a change in structure and hence a dynamic element must be inferred from the drawing (section 4.3.3). To achieve all this demands considerable facility in manipulating structures. Above all it is essential to appreciate what the drawings themselves mean, what data they are based upon and what assumptions are embodied in them. Misunderstandings and misinterpretations could lead not only to erroneous conclusions but more seriously to the breakdown of the system of communication through structural formulae.

4.3.3 *Dynamic elements in structural formulae*

A chemical reaction is not a static state but is a dynamic process; a redistribution of electrons occurs in the making and breaking of bonds. Normally the equation for a reaction expresses only the initial and final states, using the appropriate structural formulae. Unless some dynamic element is introduced into the notation the equation is silent on how the reacting groups interact. In reactions that can be envisaged as movements of electron pairs, this dynamic element is conventionally expressed by the most successful fiction of the conceptual model, the curly arrow. Beginners in mechanistic chemistry often find this representation difficult to understand and to use. Perhaps this is because they do not appreciate that the curly arrow is only the symbol of a conceptual dynamic link between structural formulae. The effective use of curly arrows, like that of any other symbol, needs practice.

Most readers will probably have experience of the curly arrow and its use, but before we indulge in exemplification and criticism it will pay to remind ourselves of what a curly arrow means. Surprisingly, few elementary text-books take the trouble to explain. They seem to expect the reader simply to absorb the meaning from the context and gist of the discussion. This is usually not difficult for simple reactions but the real danger lies in the fact that the students are not made aware of the limitations of the curly arrow, which is particularly regrettable in view of its widespread use and power.

A curly arrow represents the formal movement of an electron pair from the tail of the arrow to the atom or bond contacted by the arrowhead. Atoms so connected take a share in the electron pair and become covalently bonded. Thus the reaction of an alkyl halide with an alkoxide may be represented conveniently by

$$R^1O^- + R^2.CH_2{-}X \longrightarrow R^2.CH_2{-}OR^1 + X^-.$$

Arrow (*a*) represents the nucleophilic attack by alkoxide on the alkyl halide and arrow (*b*) the breaking of the C–X bond, X^- departing with the electron pair.

One limitation of this system is obvious. Talking about electron pair bonds is the language of VB theory and therefore curly arrows must operate upon VB structural formulae. We shall have more to say about this in a moment. Such limitations conceal dangers. One of the most common errors among beginners is the failure to draw arrows precisely from the origin of the electron pair to its exact destination, a failing often aggravated by a poor structural formula drawing. This error always leads to confusion and frequently to mistaken conclusions. It is a real weakness of the curly arrow model that it can be so easily misused – careful drawing is to be encouraged.

If these elementary difficulties are overcome, the curly arrow model performs an extremely valuable service to chemists in both analytical and creative modes. The analytical facets can be illustrated by the rationalisation of a complex rearrangement of the familiar tetracycline molecule shown in figure 4.10.

In this rearrangement, the abstraction of a proton from the tertiary alcohol by the base hydroxide ion initiates the reaction. This step is indicated by curly arrow (*a*). Arrow (*b*) shows how the resulting alkoxide anion can add into the positively polarised end of a carbonyl group to give the intermediate bridged compound (4.10.1). The strain in this bridged intermediate is removed by the reforming of the carbonyl group and the cleavage of the adjacent carbon–carbon bond as shown by arrows (*c*). Such sequences of electron shifts are commonly dignified with the title of 'mechanism' even when scant experimental evidence is available to support them. Provided that it is recognised that such schemes represent nothing more than a useful rationalisa-

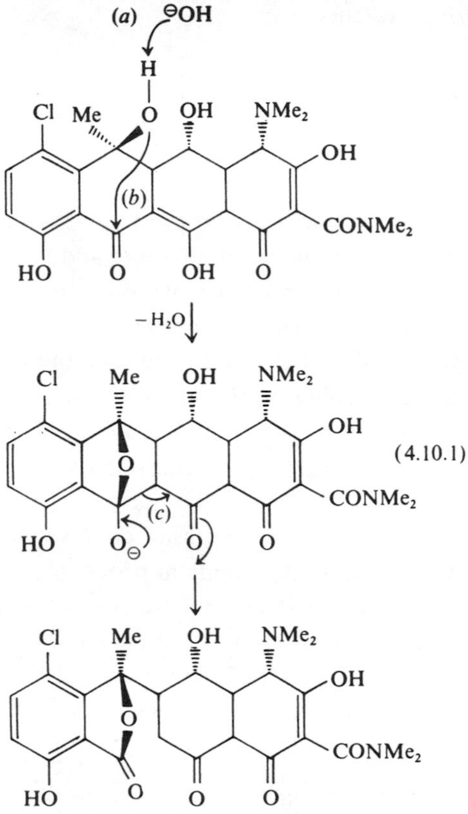

Figure 4.10. A rearrangement reaction of chlorotetracycline. The arrows marked on (4.10:1) lead to an enol which tautomerises to the product shown.

tion of a reaction, then it is reasonable to accord the description the title mechanism.

The curly arrow model assessed. Too much significance must not be attached to the arrows themselves because one cannot be sure which electrons move when. It is particularly difficult to apply arrows to an MO formulation of a reaction. Arrows are perhaps best thought of as a fragile bridge that spans the unknown space between defined structures but which may well carry the mind to firmer ground that can be consolidated experimentally.

Despite these limitations, curly arrows have the virtue of introducing a dynamic effect into the conceptual model and do so in a way that is, at least as far as reactants and products are concerned, chemically reasonable, and of allowing intermediates to be predicted for complex reactions. Their countervailing vice is that they suit the conventional conceptual model so well that any reaction mechanism that is advanced with the embellishment of curly arrows can be over-eagerly accepted as fact.

Nevertheless, provided that we are careful not to endow them with too much physical significance, arrows denoting one and two electron shifts are useful in representing in an easily visualised way the bond making and bond breaking processes in a reaction. Their use ensures that the intermediates proposed and the end products themselves are chemically and electronically realistic. However the mere fact that a satisfactory theoretical mechanism can be drawn using electron shifts gives the mechanism no basis in fact – there must be supporting experimental evidence.

The use of curly arrows permits a natural, concise and easily understandable representation of some chemical processes as a series of two-electron shifts on the same structural formula and their use tends therefore to promote the postulating of concerted mechanisms because of the very simplicity and clarity of the notation. This tendency can be rationalised from a theoretical point of view as explained by Bordwell (ref. 14):

> The obvious energetic economy of a process in which bond making is aided by simultaneous and synchronous bond breaking (concerted mechanism). . .led to the rapid acceptance of the concerted mechanism by organic chemists and to the representation of the majority of organic reactions as concerted processes.

As we have mentioned, some reactions such as those initiated photochemically do not lend themselves to a VB description and hence to discussion in terms of curly arrow electron shifts. In these cases a complementary approach that has been called 'MO following' can be adopted. The intention is to try and discern what happens to the form of the molecular orbitals when they are transformed from reactant orbitals via the transition state to product orbitals. Several methods of doing this have been described and we shall return to two of them in greater detail later in this chapter (section 4.7). All of them are more cumbersome than curly arrow pushing, both in the effort required to apply them and in the knowledge of orbitals that the chemist needs. One protagonist in this field, Zimmerman, assesses his work on MO following like this:

The general approach of MO following is seen to allow
one to focus attention on the molecular orbital effects
which control the reaction energy deriving from electron
delocalisation effects. Rather than leading blindly to
predictions, MO following allows insight into the factors
controlling reaction allowedness or forbiddenness. While
really not quite as simple as electron pushing, this MO
treatment in fact parallels electron [curly arrow] pushing in
being qualitative but powerful. [Ref. 15.]

Having considered these topics, we are now able to turn to specific
reactions and to examine how they may be influenced by the environment in which the reactant molecules find themselves with the aid
of the models discussed above.

4.4 Solvation, solvent effects and salt effects

It is about as easy to run reactions in solution as it is difficult
to develop relevant satisfactory theoretical models. When a molecule, atom, or ion goes into solution the solvent molecules take notice
of the presence of a solute by aligning themselves in such a way as
to minimise the free energy of the solution. This involves a form of
chemical interaction of solvent with solute molecules that is known
as solvation. The precise nature of solvation will vary depending upon
the precise chemical characteristics of both solvent and solute, but
some change in the intermolecular structure of the solvent is inevitable. Fluoride ions in liquid hydrogen fluoride, and toluene in benzene
show the extremes of solvation: the former involves some of the
strongest hydrogen bonds known and the latter relatively weak
'hydrophobic' interactions. Although solvation is itself in fact a type
of chemical reaction, it is normally discussed separately from the
other reactions that may occur in a given system.

The extent of solvation of a given solute varies from one solvent
to another and if a reaction rate or a reaction course is changed by
carrying out the reaction in a different solvent we speak of the operation
of a solvent effect. Frequently, the ability to perform a desired reaction
depends critically upon the choice of solvent.

4.4.1 Solvation

The visible difference between cobalt(II) chloride in water, in
which it forms a pink solution, and in methanol, when it is blue, is but

one example of the striking effects a solvent can have upon a solute. The effects are rarely so easily observed as this colour difference but they can have substantial importance even in simple chemistry. As an example, consider the relative basicities of the following series of amines measured in a polar solvent, water.

Amine	NH_3	$MeNH_2$	Me_2NH	Me_3N
pK_a	9.25	10.64	10.77	9.80

On the basis of the inductive effect of the methyl group (electron donating) we would clearly expect the tertiary amine, Me_3N, to be the most basic of the four but, as the figures show, it is only a little stronger than ammonia. This order of basicity is found in most polar solvents and does not agree with the predictions of the conceptual model if solvent effects are neglected. However when the basicities were measured in the gas phase where solvation is absent, the observed values followed precisely the order expected from the operation of inductive effects (ref. 16). The discrepancies found in solution can therefore be ascribed to difference in the solvation of the various classes of amines and their conjugate acids. The significant difference is in the size and shape of the amines which affect solvation, although the detailed relative orientations of solvent and solute molecules are unknown.

In general, solvation can be considered as an ion–dipole or dipole–dipole interaction which, as would be expected, leads to the association of oppositely charged species. In hydroxylic solvents, in amines and in hydrogen fluoride there exists in addition the possibility of hydrogen bonding between the solvent and the solute. All such inter-actions change the energy of the system as a whole and this energy change, the solvation energy, can be measured calorimetrically.

Solvation can also be considered in terms of the HSAB model (chapter 3); almost all metal ions in the gas phase are hard. This is because the polarising influences of solvent dipoles are absent in the gas phase and is consistent with the observation that solute ions show harder behaviour in solvents of lower dielectric constant (ref. 17).

In the case of the amines, it is difficult to picture the structural features that underlie the differences in solvation because the solvated molecules are in effect only differently composed dipoles. However the situation is quite different in the case of transition metal ions. A characteristic of transition metal ion chemistry is the dependence of the colour of a compound (or more precisely the dependence of its electronic spectrum) upon the geometry and coordination of the compound (chapter 3). Consequently it is reasonable to suppose that the colours of transition metal ions in solution reflect the coordination of the solvent molecules, that is to say their solvation. Ideas such as this grew with the development of ligand field theory during the 1930s and the experimental evidence that was gathered amply supported the theories. Perhaps the most interesting outcome from the point of view of a unified model of chemistry is that such observations provide compelling evidence for the reality of the intuitive notion of ion atmospheres which is a part of the Debye–Huckel–Onsager theory of electrolytes (chapter 3). In contrast to dilute solutions of electrolytes, in which ions can be regarded as spheres, we are now concerned with a directional interaction of a solvent and an ion. There is a precise directional component rather than an ion atmosphere.

In cobalt chloride, the pink colour found in aqueous solution is characteristic of the monomeric ion $Co(H_2O)_6^{2+}$ which is octahedral whereas the blue colour (in methanol) characterises a polymeric, tetrahedrally coordinated $(CoCl_2)_n$ species which is found not only in solvents of low dielectric constant but also in the gas phase in the mass spectrometer (ref. 18). It is possible to observe such specific solvation directly using X-ray diffraction techniques. In this way cobalt(II) bromide has been shown to exist predominantly as $Co(H_2O)_6^{2+}$ in aqueous environments but as $CoBr_2(MeOH)_2$ in methanol (ref. 19). Clearly one would not expect identical chemistry to occur in both solvents since different chemical species are present in each.

By considering solvation to be as specific as coordination in a crystal, however, we are hiding an important aspect of the phenomenon. Solvation is a dynamic process in which the solvent molecules which are at any instant coordinated to the solute ion undergo continuous exchange with the molecules of the bulk solvent. In suitable cases where the rate of exchange is close to the usual nmr time scale it is possible to demonstrate the occurrence of this exchange directly. This is because molecules of the bulk solvent have a different chemical environment and hence a different chemical shift from those molecules which are coordinated. In water such experiments have been done using the resonance of the ^{17}O nucleus (ref. 18). Two clear ^{17}O resonances were observed. Such observations as these also support the concepts of coordination spheres and ion atmospheres.

4.4.2 *Solvent and salt effects*

Solvent effects in substitution reactions. The compelling evidence that we have just described leads us to ask how solvation may affect a reaction. The most clear cut examples of such effects arise when the quantitative stretches into the qualitative and the solvent becomes so different that not only is the reaction rate changed but a new mechanism is brought into play. A very sensitive indicator of reaction mechanism applicable to organic, inorganic and biological chemistry is the stereochemical course. Some of the earliest stereochemical studies of reaction mechanisms were carried out by Hughes (ref. 20) who observed the simultaneous racemisation and incorporation of radioactive iodine into 2-octyl iodide on treatment with radioactive iodide in acetone (4.11.1). The reaction, a substitution process albeit of one iodine atom for another, was found to be bimolecular and the mechanism of the reaction was dubbed S_N2 which is shorthand for substitution nucleophilic bimolecular.

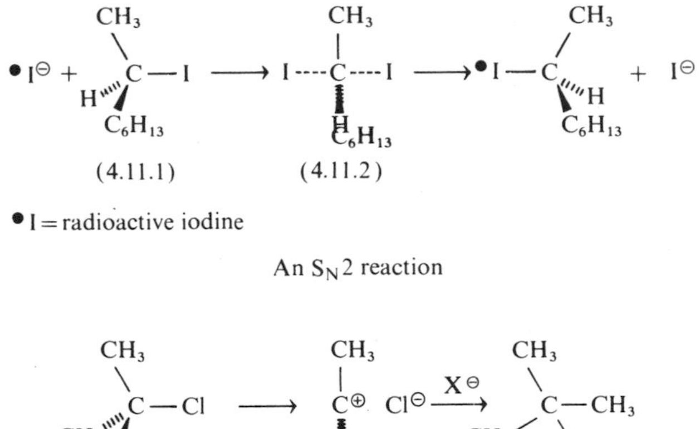

(4.11.1)   (4.11.2)

• I = radioactive iodine

An S_N2 reaction

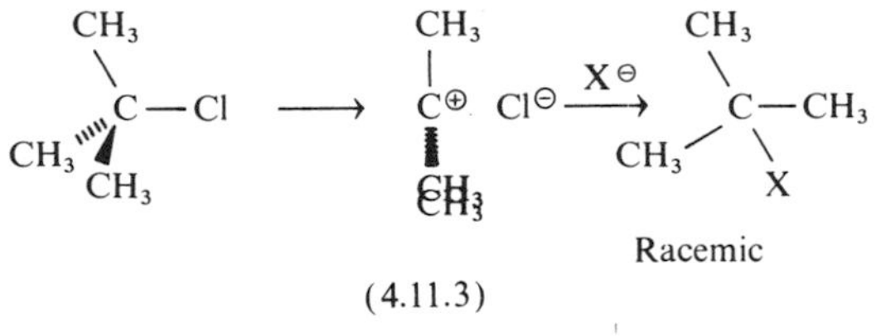

(4.11.3)

An S_N1 reaction

Figure 4.11.

Hughes and Ingold recognised that this mechanism was chiefly characteristic of primary alkyl halides and that inversion of configuration occurs in the course of the substitution. Other characteristics associated with the S_N2 label are an insensitivity of the reaction to changes in the solvent (see below), and an uncharged transition state. A pictorial representation of the transition state for an S_N2 reaction is difficult to draw using conventional formulae since it contains one half-made and one half-broken bond. The usual representation is shown in (4.11.2) in which the dotted lines imply a partial bond. The value of such notation has been questioned but MO calculations have led Allinger to conclude that this is the best simple way of expressing the S_N2 transition state diagrammatically (ref. 21).

Hughes and Ingold found that the S_N2 mechanism was only one of two major classes of mechanism for nucleophilic substitution reactions in alkyl halides. The other was a unimolecular process in which the alkyl halide initially dissociates to form a carbenium ion and a halogen anion. The carbenium ion is rapidly captured by the attacking nucleophile (X^- in 4.11.3). This mechanism was termed S_N1 (short for substitution nucleophilic unimolecular) and is characteristic of tertiary halides.

Solvents will influence these reactions most at the moments when charge is established or removed and Hughes and Ingold identified these moments as the transition state for S_N2 reaction and the carbenium ion for the S_N1 reaction. They therefore proposed (ref. 22) that 'an increase in solvating power will accelerate the creation and concentration of charges and inhibit their destruction and diffusion'. This qualitative rule of thumb was complementary to ideas which had been put forward by Brønsted and Bjerrum to describe quantitatively the effect of adding non-reactant ions to a reaction medium. A change in the ionic strength of the medium causes a change in the structure of the solvent owing to the solvation of the added ions and hence the reaction environment is modified. A change in the rate of a reaction when ions are added is known as a salt effect.

A quantitative model of salt effects. Brønsted and Bjerrum developed their theoretical model of salt effects by integrating two well-established theoretical models of physical chemistry, the Debye–Huckel–Onsager theory (chapter 3) and transition state theory (section 4.2.2). They derived the equation 4.7 for the reaction of ions A and B (ref. 1e)

$$\log \frac{k}{k_0} = 2z_A z_B \, \alpha \mu^{\frac{1}{2}} \qquad (4.7)$$

where k is the rate constant for the reaction in question, k_0 is the rate constant for the reference reaction, z_A and z_B are the charges on ions A and B, α is the Debye–Huckel–Onsager constant, and μ is the ionic strength of the solution.

This equation implies that an increase in ionic strength causes an increase in the rate of reactions involving like charges and vice-versa: the result is compatible with Hughes' and Ingold's ideas about solvent effects because solvation is tantamount to the stabilising of an ion by surrounding it with ions (or poles of dipoles) of the opposite charge.

A quantitative model of solvent effects and its use in mechanistic studies. To return to the mechanism of substitution reactions, we implied earlier that the transition state for an S_N2 mechanism, although more charged than the reactants or products, is less so than the intermediate or transition state for an S_N1 process which should resemble a carbenium ion. We would expect, therefore, that the rate of an S_N1 reaction would be more sensitive to solvent effects than the rate of an S_N2 reaction. For reactions that proceed via an S_N1 mechanism, the variation of rate with solvent can be correlated with a parameter Y that measures the ionising power of the solvent. This has been done by Grunwald and Winstein (ref. 23) by means of equation 4.8

$$\log \frac{k}{k_0} = mY \tag{4.8}$$

where k and k_0 have their usual significance and m is a constant characteristic of the compound undergoing substitution.

Models of this type can be used in two general ways – either in structuring the results of an investigation into the properties of a particular solvent, or, by examining one reaction in a number of solvents, they can be an aid to defining a mechanism. A reaction that occurs by an S_N1 mechanism would yield rate constants which could be fitted into equation 4.8 with one value of m, as would be expected from the definition of m embodied in the equation. Such correlations apply equally to organic and inorganic substitution reactions (ref. 24). For example, the aquation of chloropentammine cobalt(II) ion $[Co(NH_3)_5Cl]^+$, in aqueous perchloric acid shows a decrease in rate that correlates with decreasing solvating power (dielectric constant) of the reaction medium. Such behaviour is consistent with a dissociative mechanism as illustrated in figure 4.12.

$$[Co(NH_3)_5Cl]^+ \xrightarrow{\text{slow}} [Co(NH_3)_5]^{2+} + Cl^- \xrightarrow[H_2O]{\text{fast}} [Co(NH_3)_5(H_2O)]^{2+}$$

Figure 4.12

Relationships between the models of solvent and salt effects. We have just traced links between three independently deduced models of solvent and salt effects, namely the Hughes–Ingold qualitative rule, the Brønsted–Bjerrum model, and the Grunwald–Winstein model, and in the process we saw the interdependence of theoretical and empirical models in chemistry. This is a theme of this book. If we look more closely at the last two models, we find also a close parallel to the modelling procedure employed by Hammett in correlation of substituent effects in acid–base equilibria (chapter 3). The parallel lies in the use of a reference or model reaction and thus the caveats that applied to Hammett's work and to linear free energy relationships apply equally here.

4.5 Catalysis as an environmental effect

Another way of describing the effect of solvents on reactions is through the effect of the solvent upon the transition state. If the transition state energy is lowered by the solvent (as in the S_N2 reaction in a non-polar solvent) then the reaction rate will be increased. This environmental effect brought about by the solvent can be generalised to describe the extremely important phenomenon of catalysis.

Catalysis is caused by changing the environment of a reaction in such a way that the reaction rate is increased through the mediation of a catalyst which undergoes no net change in the course of the reaction. There are two distinct ways in which this can be done. Either a new component, the catalyst, is introduced into the same phase as the reaction (homogeneous catalysis), or a new phase which is a catalyst is added (heterogeneous catalysis). The heterogeneous case is an obvious change of environment in the most general sense of the word but, remembering that all components outside the molecule of reactant constitute that molecule's external environment, we must describe homogeneous catalysis as an environmental effect too. When we come to consider catalysis in the context of biochemistry in chapter 6, the importance of the environment that the biological catalysts, enzymes, provide for a reaction becomes very significant. As a foretaste of this, let us consider some simple chemical examples in which catalysis produces a favourable environment for a reaction.

The environment in homogeneous catalysis. Examples of homogeneous catalysis in the laboratory readily come to mind. The proton, for example, is involved in the catalysis of ester formation and hydrolysis. In HSAB terms, the proton is the hardest of all ions since the positive charge is totally unshielded by a polarisable electron cloud. Therefore the proton is the most potent catalyst where a bond is to be made susceptible to nucleophilic attack. The formation of acetals (figure 4.13) is catalysed by protons. Protonation of the ketone yields an oxonium ion (4.13.1) which will be much more readily attacked by the

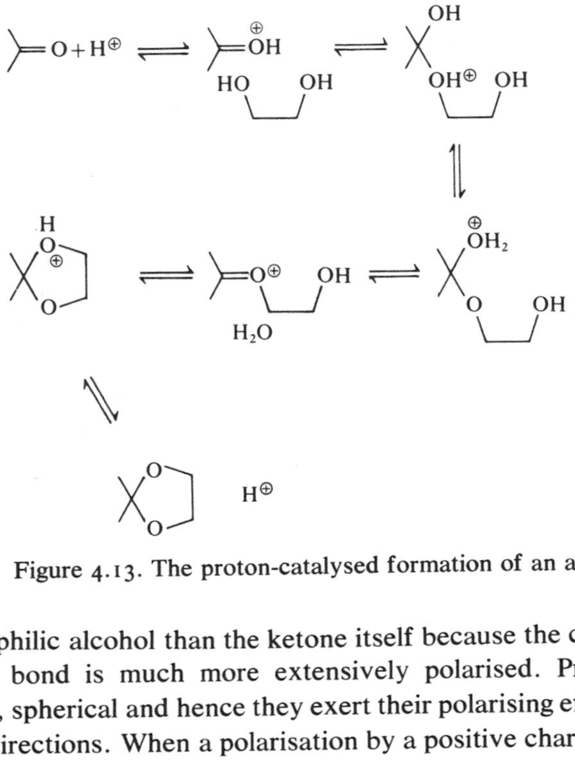

Figure 4.13. The proton-catalysed formation of an acetal.

nucleophilic alcohol than the ketone itself because the carbon–oxygen double bond is much more extensively polarised. Protons are, of course, spherical and hence they exert their polarising effect uniformly in all directions. When a polarisation by a positive charge in a specific direction is required then a metal ion is appropriate. We shall meet an example of this with the mechanism of action of the enzyme carboxypeptidase (chapter 6).

The preceding example illustrates how a catalyst can combine with the substrate to form a more reactive intermediate than would otherwise be available. The same effect occurs in a more subtle manner when the catalyst acts as a carrier as in the redox reactions of some transition metal complexes (ref. 25). Compare the relative rates of the following two reactions:

$$Co(NH_3)_6^{3+} + Cr^{2+} \xrightarrow{\text{relative rate } k = 1} Co^{2+} + Cr^{3+} + 6NH_3$$
$$Co(NH_3)_5Cl^{2+} + Cr^{2+} \xrightarrow{k = 6.10^9} Co^{2+} + CrCl^{2+} + 5NH_3.$$

This difference in rate can be understood if the chloride ion present in the chloropentammine complex acts as a bridge between the two metal ions and aids the transfer of electrons. Many organic carboxylate anions as well as chloride can act as bridges in such redox reactions.

In the previous sections, we developed the idea that a change of solvent could promote one reaction at the expense of another. Looking at it the other way round, we could say that the change in environment is suppressing an alternative reaction pathway. Such negative catalysis, or inhibition, is most simply illustrated by non-polar solvents which, as we saw, cannot stabilise ions by solvation. We shall see in chapter 6 how this effect allows the respiratory proteins to function and how models have been used to demonstrate the chemistry behind their behaviour. Similar environmental effects can be provided by aromatic hydrocarbons and long chain aliphatic compounds which, because of their non-polar nature, exclude water and other polar groups from their vicinity. This causes polar compounds, in turn, to group themselves away from non-polar regions. The ordering of polar and non-polar molecules is associated with a favourable free-energy change and under appropriate circumstances, the association of non-polar groups (so-called hydrophobic bonding) can lead to potent catalysis (ref. 26).

Here, then, are cases in which catalysts provide a qualitative change in environment. Of course it is desirable to have quantitative models of catalysis and it is interesting to see that the models used are closely similar in form to examples of models that we have already discussed in other contexts.

A linear free energy model of acid–base catalysis. Catalysis always involves a change of mechanism and there are many theoretical models that express this. For instance Brønsted introduced linear free energy relationships to correlate data collected from acid–base catalysed reactions. The usual correlation procedure was applied; the reaction in question is described in terms of a linear correlation with a reference reaction and a property associated with the type of catalysis that is occurring. Mathematically this takes the form of equation 4.9

$$\log \frac{k_A}{G_A} = -\alpha p K_A.$$ (4.9)

Here, k_A represents the rate constant for catalysis by acid A of acidity pK_A, G_A is a constant for the reaction in question (the reference point) and α is the parameter that describes the type of catalysis that is occurring. Many values of α have been reported for reactions in which the catalytic mechanism could be determined by other means: an experimentally determined α can therefore be used as a criterion of mechanism (ref. 4b). The interested reader can find much more information about homogeneous catalysis in solution and about the application of correlations of this type in two excellent books (refs. 4a and 27).

Heterogeneous catalysis. Heterogeneous catalysis is more difficult to characterise than homogeneous catalysis because the physical nature of the catalyst surface is important and we have, so far, no language to describe it. Such surface phenomena require specialised theoretical models and they are at the limit of that which current conceptual models can interpret. In certain cases, some insight into the mechanism of heterogeneous catalysis at the molecular level can be gained from symmetry rules (section 4.7). Recently some models of heterogeneous oxidation catalysts have been described (ref. 28). They draw a parallel between reaction sites on the catalyst surface and functional groups in a molecule; the model has the same structure as our description of homogeneous reactions in general.

4.6 Models of transition states

The possibility of stabilising a transition state by catalysis points to the need for a conceptual description of transition states both in qualitative and quantitative terms. The difficulty is that the transition state is a maximum on the energy profile, it is a transient species whose properties we cannot measure directly. Hence the usual control of comparing a model's predictions with experimental results are less reliable because, in order to relate observation to calculation, a further set of transformations must be effected.

4.6.1 *Empirical and qualitative models*

Hammond's hypothesis. At a pictorial level the transformations which relate the prototype transition states to their models are so vague that it is pointless to worry about detail. This is especially so in a number of cases that can be represented by the energy profiles shown in figure 4.14, some of which we have already met. Hammond (ref. 29) realised that the energy profiles point to the nature of the transition state and he suggested that in a strongly exothermic reaction, the transition state must resemble the reactants (4.14.1). Conversely, in an endothermic process, a product-like transition state is probable (4.14.2). If a metastable intermediate is formed (4.14.3), we can argue from the experimentally proven structures of such species and describe the transition state as being closely related to the intermediate. For this reason we were able to assert earlier that the transition state for an S_N1 reaction must be more polar than that for an S_N2 reaction (section 4.4.2). Hammond's hypothesis is a useful supplement to the dynamic elements of the conceptual model because it provides some basis for understanding the probable structure of a transition state.

Clearly a knowledge of the structure of the starting materials and products will always allow us to propose hypothetical transition states and, if a series of related reactions show regularities, it may be possible to systematise the available information in a model from which predictions can be made. Tacitly this assumes a constant relationship between the reactants, transition state and products and amounts to no more than a none too robust analogical process.

Cram's rule. A quite complex but effective method for comparing the energies of competing transition states is known as Cram's rule (ref. 30). It applies to nucleophilic additions to aldehydes and ketones that have an asymmetric centre at the carbon α to the carbonyl group. The results of a typical experiment are shown in figure 4.15. Cram's model of the transition states is used to interpret the results as follows. The substrate, an asymmetric carbonyl compound, can be drawn as (4.15.1) in which the substituent alkyl groups are classified as small (S), medium (M), or large (L). For the case of reduction of aldehydes and ketones by metal hydrides the group R can be a hydrogen atom or an alkyl group, Cram proposed that the configuration of the transition state which has lowest energy (that is to say the kinetically favoured transition state) is such that the hydride ion attacks the face of the

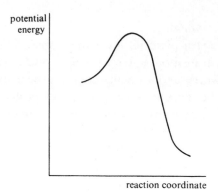

(4.14.1) An exothermic reaction

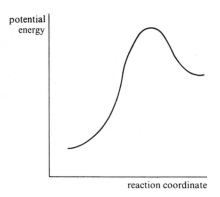

(4.14.2) An endothermic reaction

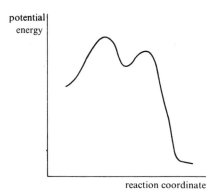

(4.14.3) A reaction involving a
metastable intermediate

Figure 4.14

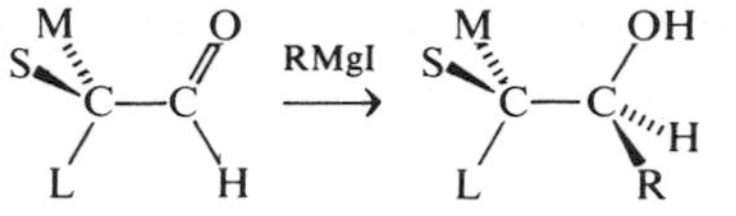

(4.15.1) (4.15.3)

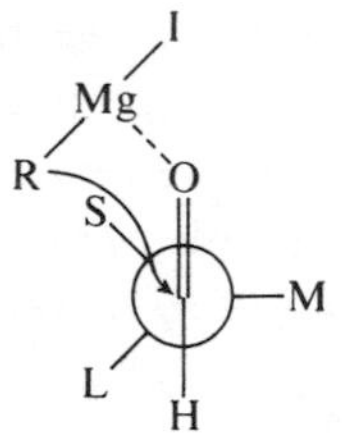

(4.15.2)

Figure 4.15. An example of the use of Cram's rule.

carbonyl group from the least hindered side. This side is expected to be adjacent to the smallest substituent S, as indicated by the arrow. Thus the transition state can be envisaged as (4.15.2) and the products will be expected to be (4.15.3). Once again we see how the transition state determines the products that are formed initially. However, subsequent reactions can lead to equilibration with a more thermodynamically favoured isomer and the rule may in these cases apparently fail.

Limitations of models like Cram's rule. Rationalisations like Cram's rule have been widely applied with considerable success. Their virtue is their simplicity of approach which limits the system to only one variable factor, the steric effect which is deemed to be of prime importance in determining the configuration of the transition state. This model transition state (4.15.2) is only a generalised representation: the rule makes no attempt to predict a precise transition state structure in the way that theoretical models have done for reactions such as nucleophilic substitution as we shall see shortly (section 4.6.3). In addition many features that may, *a priori*, be of importance are omitted from the model. These include coordination of the substrate with the reducing agent through the metal, solvent participation and thermodynamic control of the reaction. Consequently, Cram's rule is restricted to reactions in which these factors are unimportant. Within this range, the rule successfully predicts the major products observed and has been extended, perhaps inadvisedly, to a semi-quantitative level. Fortunately chemists have realised the severe constraints which exist in the application of such models of transition states and Morrison has commented

> It is important that one realise the empirical nature of all such models and appreciate their possible fallibility and theoretical naivety... These transition state models can be useful conceptual devices for correlating experimental results but they should not be given credence as predictive tools unless a large number of examples has been successfully accommodated. The utility of the model for correlative and predictive purposes does not necessarily confirm it as an accurate representation of the product controlling transition state. [Ref. 31]

The same could be said about the quantitative approaches which attempt a more fundamental understanding of transition states, as the following sections show.

4.6.2 *Quantum mechanical calculations of transition states*

In principle it is possible to calculate the energy of a molecule using quantum mechanics with a knowledge only of the constitution of that molecule. We saw, however, that this is difficult for all but the simplest molecules and it is not surprising that application of the methods to transition states, which are themselves poorly defined entities, calls for a higher degree of empiricism than do calculations of ground states. Empiricism can be enlightened by intuitive and experimental understanding of the chemistry behind the reaction and of the transition state itself. Consequently, some of the most successful approaches to the calculation of the properties of transition states have been semi-empirical, a threefold compromise between rigour, experiment and intuition.

Most methods are MO based. One of the most frequently used is due to Pople (ref. 32) and it is known as the CNDO method. As is common practice in quantum mechanical theories, the name, CNDO, stands for a major feature of the mathematical model being used. In this case, Pople makes the assumption that it is permissible to Completely Neglect Differential Overlap of orbitals. As one might expect, models that embody a drastic assumption are susceptible to refinement and thus a further generation of models is begotten. We shall now discuss some of the results of the MINDO (Modified Incom-

plete Neglect of Differential Overlap) methods developed by Dewar
(refs. 33, 34). These models are related to Pople's CNDO method and
have now reached the third generation (MINDO/3).

The MINDO method is a semi-empirical approach. It begins from
the exact Schrödinger equation for the system in question and replaces
some of the integrals required to solve the equation by parameters
which are evaluated from reference compounds and with the aid of
atomic spectroscopy. Precisely this modelling procedure can be used
for calculations of ground states also and once again the chief problem
is the evaluation of the empirical parameters. Dewar, well aware of
the difficulty, has commented that there is no point in having equations
of the right form if they give meaningless quantitative predictions. He
has tested the MINDO methods with many hundreds of compounds and
the following data give some indication of the accuracy possible for
ground state molecules. The same parameters were used for all the
compounds.

Predictions of MINDO calculations

Compound	Heat of formation (kJ mol^{-1})	
	Calculated	Observed
Stable compounds[a]		
C_2H_6	−90.8	−84.5
C_2H_4	68.6	52.3
C_2H_2	223.4	227.2
azulene	282.0	288.3
Ions[b]		
CH_3^+	1089.8	1088.6
$(CH_3)_2CH^+$	780.4	799.5
allyl$^+$	930.3	904.4, 946.2
Reactive intermediates[b]		
singlet methylene	422.8	419.5
triplet methylene	383.1	393.1, 389

(a) Ref. 34 (MINDO/3).   (b) Ref. 35 (MINDO/2).

A theoretical treatment that stands up so well to the test of experiment for stable compounds, ions, and reactive intermediates is
justifiably a candidate for application to the thorny problem of
calculating the structure and energies of transition states.

If it is desired to calculate what the transition state for a reaction is likely to be, it is obviously a mistake to assume what the geometry of the transition state is, for to do so would prejudge the mechanism of the reaction. MINDO makes no such assumptions and, bearing in mind the reliability demonstrated for ground state molecules, the predictions of MINDO should be taken seriously.

The transition state can be located by finding the minimum energy reaction path; the point of maximum energy on this path corresponds to the transition state. This has been done for the Cope rearrangement of 1,5-hexadiene (figure 4.16). There are two likely geometries for this reaction, via a chair-like species (4.16.1) and a boat-like one (4.16.2). MINDO calculates that the symmetrical chair and boat structures illustrated are not the transition states but are metastable intermediates, minima on the reaction profile rather than maxima. The transition state, as is often found in such calculations, is asymmetrical. Also MINDO predicts in line with intuition that the chair path will be of lower energy than the boat path, although the numerical errors are larger than for ground state molecules.

| | Activation energy (kJ mol^{-1}) | |
	Observed	Calculated
Chair path	140.3±2	140.3
Boat path	187.1±8	166.6

Similarly the conversion of Dewar benzene (4.16.3) to benzene has the calculated activation energy of 114.3 kJ mol^{-1} compared with the experimentally determined value 96.3 kJ mol^{-1}, and once again the transition state is asymmetrical. It may be that such calculations will provide as much information as can be obtained about transition states which current experimental techniques are incapable of observing.

Although MINDO/3 is parametrised for ground state molecules, it also gives reasonable results for first excited states. Chemiluminescence, for example, can arise in the thermolysis of oxetanes (4.16.4). This reaction produces a ketone in an excited state, the decay of which emits light. For the simplest oxetane (4.16.4, R = H), the calculated triplet excited state energy of the excited ketone is 305.2 kJ mol^{-1}, and 301.5 kJ mol^{-1} was observed.

The above calculations have all been checked by experiment. How-

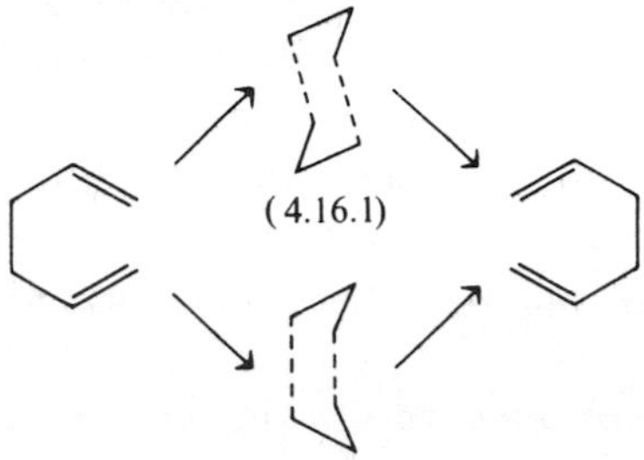

(4.16.1)

(4.16.2)

Cope rearrangement

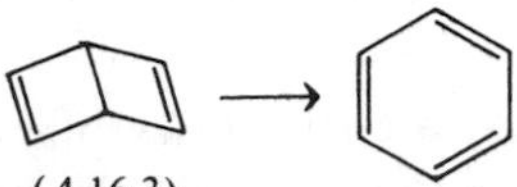

(4.16.3)

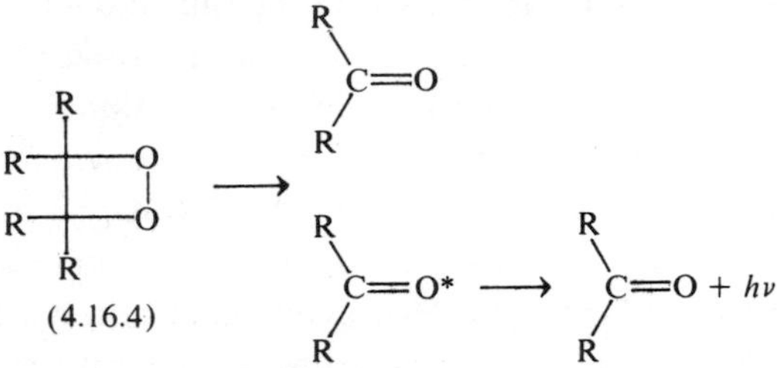

(4.16.4)

Excited ketone

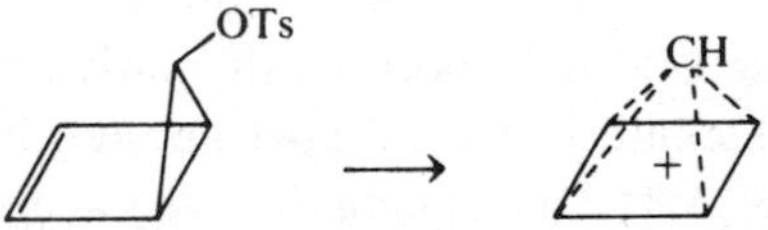

(4.16.5) → (4.16.6)

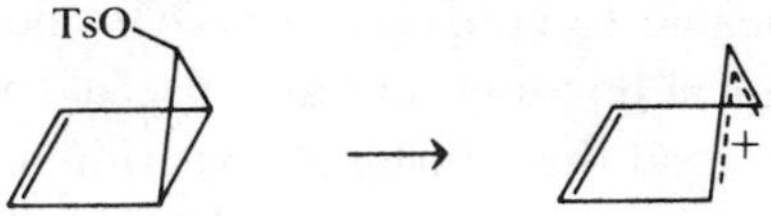

(4.16.7) → (4.16.8)

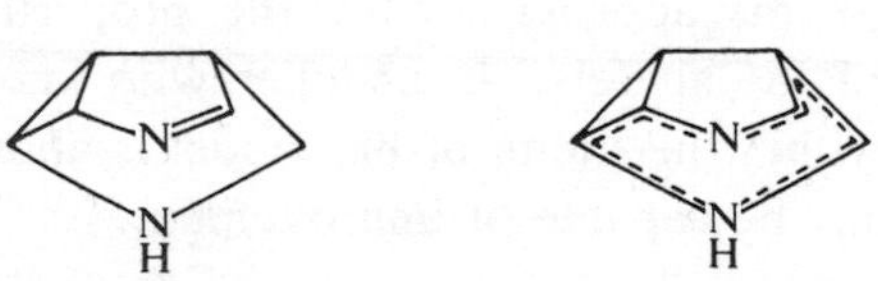

(4.16.9)   (4.16.10)

Figure 4.16

ever MINDO/3 has also made some interesting predictions yet to be tested by experiment. For example, on solvolysis of bicyclo-(2,1,0)-pentenyl esters MINDO predicts that the anti isomer (4.16.5) will give a square pyramidal carbonium ion (4.16.6) whereas the syn isomer (4.16.7) should give the allyl cyclopentenium ion (4.16.8) (ref. 34).

As if to emphasise the problem of drawing conventional structures for very delocalised molecules, MINDO points out that the 'classical' diagram (4.16.9) is not the best representation of the diazabullvalene. Better is the drawing (4.16.10) which uses the dotted line to indicate the delocalisation of electrons over all the atoms that it connects (ref. 36).

The foregoing examples illustrate the versatility of MINDO/3 in the range of reactions of interest to the practical chemist that it can treat. One of Dewar's aims in constructing this model was to supply a theoretical treatment that could be applied to problems of general chemical interest with a minimum of computational expense. In building this constraint into MINDO/3 some accuracy must be sacrificed and there are several systems for which MINDO/3 is inadequate. These were recognised by Dewar (ref. 37) and have been underlined by Pople (ref. 38) and Hehre (ref. 39). However, real as they are, the limitations of MINDO/3 as argued by these authors concern reactions of a more 'academic' character, of interest more to the theoretician than to his practical colleague. *Ab initio* calculations in some cases do give better results but at the expense of much more mathematical labour. Dewar estimates that such a calculation would cost about 300,000 times as much as the comparable MINDO calculation for a molecule of the complexity of the norbornyl cation.

Hence it seems reasonable to consider MINDO/3 not as a general universal model of chemical bonding, but as a useful tool for investigating reactions on or beyond the fringes of experiment. In this more limited way, MINDO/3 or its successors may have great value in the future. Theoretical chemists may object but the description 'quantitative curly arrow' seems appropriate for the properties of semiempirical MO methods as models. A modeller who knowingly confines his operations within the limits of his model is above criticism, although his model may be capable of improvement.

4.6.3 *Quantitative methods based upon classical mechanics*

One of the most difficult parameters to estimate by semi-empirical methods is the effect of strain, although MINDO fares better than most. This has, in fact, been most successfully achieved using a theoretical model based upon classical mechanical methods that are a logical extension of the studies of Westheimer and others (section 3.2.2). Once again the choice of suitable reference compounds is important and, within this usual limitation, the classical mechanical methods have been applied with success to the calculation of vibrational spectra and thermodynamic properties of reactions involving hydrocarbons, cyclic ketones, polymers, and peptides (ref. 40).

One may well ask why one should resort to a classical treatment when satisfactory results may be available from more rigorous quantum mechanical theories. In general, classical models are easier to visualise in pictorial terms and are much easier to operate than quantum mechanical models. Less computation is required and accordingly larger molecules can be studied by classical methods. As Allinger explains, showing a thorough appreciation of models (ref. 41):

> The mechanical model which we use to represent a molecule is at present very crude in comparison to the complex elegance actually dictated by the Schrödinger equation for the electronic wave function. We obviously cannot expect our present model to reproduce accurately *all* the properties of a molecule and have chosen to parametrise our model so as to fit (1) structure, and (2) energy at 25 °C and are prepared to sacrifice accuracy for other quantities to some extent as necessary.

Classical models and substitution reactions. It is of obvious interest to compare the results of theoretical treatments with experimental correlations for particular types of reaction. Nucleophilic substitution, a reaction that has held the interest of physical organic chemists for many years and for which data is consequently rich, offers a particularly favourable case for study.

Bingham and Schleyer took up the challenge and studied the reaction illustrated in figure 4.17 (ref. 42), in which the carbon atom at which substitution occurs is at a bridgehead, Nu⁻ represents a nucleophile, and X a leaving group. Systems typical of this general reaction scheme

(4.17.1)   (4.17.2)

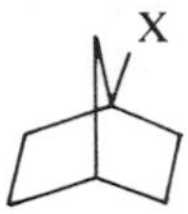

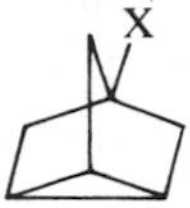

(4.17.3)   (4.17.4)   (4.17.5)

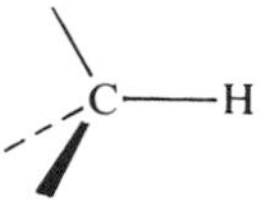

(4.17.6)

Figure 4.17

include (4.17.3–5). By choosing bridgehead systems for their calcula- ·
tions, Bingham and Schleyer greatly simplified their task because the
mechanics of these molecules prevents twisting and bending motions,
such as would be found in acylic halides. Also the solvent cannot
participate either by nucleophilic substitution or by elimination because the ring systems block its approach to the reaction site. The only
possible reaction is an S_N1 process in which the formation of the
carbenium ion (4.17.2) is rate determining. A classical method based
upon Westheimer's equation was used to calculate the energy of the
ground state (4.17.1) and the corresponding transition state between it
and the carbenium ion (4.17.2).

To make the calculation easier it was necessary to use the parent
hydrocarbon (4.17.6), whose energy can be calculated, as a model
ground state and the transition state was modelled with the aid of
Hammond's hypothesis. In this way, the hydride ion is considered to
be a model for the leaving group X, an approximation also used by
Allinger.

The calculations that can be made using this model attempt to
estimate the difference in rate that will be observed when the various
bridgehead substitution reactions take place in compounds exemplified
by (4.17.3–5). They found that the calculated strain energies correlate
linearly with the logarithm of the rate constant for several substitution

reactions such as the acetolysis of trifluoroacetates and toluene-*p*-sulphonates. In other words the strain energy within these bridgehead esters is the determining factor for the rate of carbenium ion formation.

What is noteworthy about this model, apart from its success, is the careful definition of the system. By eliminating solvent effects and choosing compounds in which twisting and bending are impossible, attention was focussed upon the feature that the model set out to study, namely strain. In fact there is insufficient space here to describe the detailed care with which Bingham and Schleyer defined their system and the indicated limitations that their approach contained; the interested reader would do well to take a look at the original paper (ref. 42).

4.7 Models of reactions describing the relationship between reactants and products

All the models of reactions that we have described so far, although focussing attention upon the most sensitive part of the re-action, the transition state, draw heavily upon a knowledge of the properties of the reactants and products. We cannot estimate the rate of a reaction without considering the energy of the transition state, but often all the chemist wants to know is whether a reaction is feasible or not. In such 'go no-go' situations, models that relate to the inter-actions between the orbitals of the reactants to give the orbitals of the products are particularly valuable. They are based upon one of the fundamental axioms of quantum mechanics, namely that chemical bonds only form when orbitals of like symmetry and energy interact. Since a reaction involves the breaking and making of bonds it follows that the symmetry of the orbitals involved in the reaction must be compatible. In other words, to form a bond, orbitals of like symmetry are required and, on breaking a bond, orbitals of like symmetry result so that the orbital symmetry before and after reaction is conserved. This naive argument sounds plausible but, partly as a result of Dewar's work (ref. 34), it is open to question. However even if it cannot be justified, the correlation of orbital symmetry does offer a useful model for predicting reaction paths as we shall see. Recently, theories based upon these ideas have flourished and have become fully assimilated into the chemist's conceptual model along with their associated jargon.

The first comprehensive compilation of reactions controlled by symmetry accompanied by a unifying theoretical rationale was provided by Woodward and Hoffmann in their famous rules (ref. 43). They describe a variety of previously unrelated reactions that occur by way of cyclic transition states. Such reactions are termed pericyclic reactions and include the well-known Diels–Alder reaction (4.18.1–2), the Claisen rearrangement (4.18.3–4) and the dimerisation of olefins (4.18.5–6). The rules make no quantitative predictions but allow the

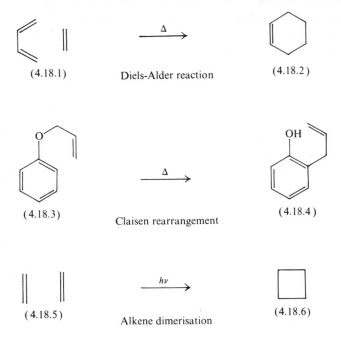

(4.18.1) Diels-Alder reaction (4.18.2)

(4.18.3) Claisen rearrangement (4.18.4)

(4.18.5) Alkene dimerisation (4.18.6)

Figure 4.18. Electrocyclic reactions.

chemist to decide which of a number of possible paths is consistent with the requirements of orbital symmetry conservation and hence will be followed by a reaction most readily.

The Woodward–Hoffmann approach. The Woodward–Hoffmann model takes the view that orbital symmetry must be conserved if a reaction is to occur with low activation energy as a concerted process. In reactions of high symmetry, it is possible to follow the change in symmetry as the reaction proceeds by means of a well-established quantum mechanical model known as a correlation diagram. Figure 4.19 shows the diagram for the dimerisation of ethylene to cyclobutane. In this reaction, two planes of symmetry are conserved and these are

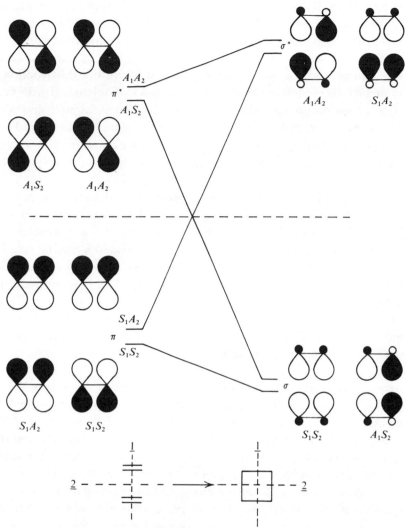

1 and 2 are symmetry planes conserved in the reaction
S_1 describes an orbital symmetrical with respect to plane 1
A_2 describes an orbital antisymmetrical with respect to plane 2 etc.

Figure 4.19. An orbital correlation diagram (adapted from
Woodward and Hoffmann, ref. 43).

labelled 1 and 2. To find out whether orbital symmetry is conserved
or not, we label each orbital as either symmetric (S) or antisymmetric
(A) with respect to the planes 1 and 2. Thus the π orbitals of ethylene
can be arranged either symmetrically with respect to just one (S_1A_2)

or to both $(S_1 S_2)$. When these orbitals have been transformed into the new σ bonds of cyclobutane, we can have either $S_1 S_2$ or $A_1 S_2$ symmetry. Thus there is a change of symmetry on reaction and this is depicted by the lines drawn linking the orbitals on figure 4.19. An analogous situation obtains for the π^* and σ^* orbitals. Therefore this reaction is symmetry forbidden in the ground state and will only occur with very high activation energy. However if an electron is excited into a π^* orbital by UV radiation, the π^* orbital, now occupied, can transform into ground state cyclobutane orbitals with conservation of symmetry $(A_1 S_2 \to A_1 S_2)$. In other words, the photochemical reaction is symmetry allowed.

These predictions agree with experiment and similar arguments apply to the other examples of figure 4.18. However it becomes very hard to draw correlation diagrams in reactions of low symmetry and consequently Woodward and Hoffman extrapolated a general rule for pericyclic reactions. Many other authors have proposed similar rules but these have been deduced from the topology of the reaction and quite independently of orbital symmetry considerations (ref. 44). These rules are necessarily terse and need practice for application. Chemists are by no means unanimous about which is the most correct theoretically or convenient practically.

Some alternative points of view. However one way of considering pericyclic reactions fits very neatly into the conceptual way of looking at things. In 1939, Evans suggested that a reaction will have a low energy pathway if it has an aromatic transition state (ref. 45), and this has been extended by Dewar (ref. 46). Aromaticity is easy to recognise by the criteria defined by Hückel's rule† (chapter 5). Thus all of the reactions of figure 4.18 can be envisaged as proceeding via a planar, cyclic transition state (they are all pericyclic) but only the first two possess the necessary 6 electrons required for an 'aromatic' transition state; the third has only four. Accordingly, the third requires photochemical activation whereas the others proceed as ground state thermal reactions. The text cited (ref. 44) provides an account of many of the models that describe pericyclic reactions.

Woodward's and Hoffmann's is not the only model possible and, as we have said, the same generalisations can be arrived at simply by considering the geometric properties of the MOs. However even these

† Hückel's rule; 'A planar cyclic conjugated system is aromatic if it contains $(4n+2)$ π electrons.'

general rules are not readily applied to reactions other than pericyclic reactions, especially those involving inorganic compounds, and there is a need for a model that will fulfil a similar function in such cases. Pearson has provided a carefully thought out model for such cases (ref. 47). Beginning from rigorous quantum mechanical equations, Pearson argues as far as he can without simplification and then carefully defines the approximations required to apply his model.

Pearson's model. Whereas Woodward and Hoffmann focus attention upon the relationship between the orbitals of reactants and products, Pearson simply concerns himself with the reactant orbitals and examines what happens to them during a reaction. The conclusion of his argument is the following criterion for a symmetry allowed reaction. If a reaction is to be allowed, there must be available low-lying excited states in one component of the reaction that can accept electrons from the highest occupied ground state orbitals of another reaction component to form a bond. In order to determine what these orbitals are, approximations are introduced.

Firstly, Pearson uses LCAO MOs because they are easiest to obtain. No serious difficulties are introduced here because one of the great virtues of MO theory is that it pays great attention to MO symmetry and Pearson's theory is based upon symmetry. Secondly, he chooses to consider only a few low-lying excited states, not all such states as the quantum mechanical calculations require. Since the argument is simply between the two extremes of allowed and forbidden, the model is also adequate in this respect. In particular, the important orbitals are identified as the highest occupied MO (HOMO) and the lowest unoccupied MO (LUMO) which were first discussed as frontier orbitals by Fukui (ref. 48) and have recently been given further significance, again by Dewar (ref. 49). The criterion can now be made more explicit. For a reaction to be allowed, electrons must be transferred from the HOMO of one component of the reaction to the LUMO of another, each orbital being of like symmetry. Electrons are transferred towards the more electronegative atom as the general conceptual description of reactions requires.

Pearson discusses many illuminating examples. For years the reaction between hydrogen molecules and iodine molecules was considered to occur by a broadside collision of a molecule of hydrogen and a molecule of iodine (section 4.2.1). If we consider the HOMO and the LUMO for the reaction, such an attack is clearly forbidden because

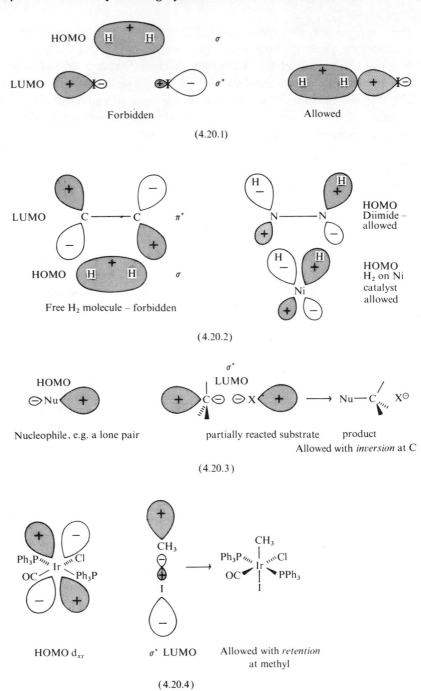

Figure 4.20. An orbital symmetry treatment of some common reaction types (adapted from Pearson, ref. 47).

these orbitals do not have the same symmetry as each other (4.20.1). On the other hand, an end-on attack of an iodine atom upon a hydrogen molecule is allowed and modern experimental evidence bears out that this is in fact the mechanism of the reaction (ref. 50).

The reduction of double bonds by hydrogen molecules is similarly forbidden (4.20.2) but a catalyst can act as a bridge between the opposite symmetries of the reactants. This could be considered as yet another example of an outer environmental effect of the type discussed in section 4.5.

For our final example, let us turn once again to the stereochemistry of substitution reactions. Pearson's method predicts that σ attack as in the S_N2 reactions of organic compounds must occur with inversion (4.20.3) but, in contrast, substitution of certain inorganic ions may occur with retention of configuration at the carbon atom. In this case π attack occurs (4.20.4).

Orbital symmetry models in perspective. There is value in all of the models that we have just discussed and it is for the chemist himself to decide which he prefers, although the last word on this topic has certainly not yet been said. It is significant that these models have surmounted the barrier of communication between theoretical and practical chemists because the concepts can be easily visualised by anyone with even a limited appreciation of MO theory. Unlike most theoretical treatments which require computer facilities, these theories can be applied non-mathematically using just pencil and paper and a clear mind. As scientific theories they have also proved remarkably successful in providing a unified conception of previously unrelated facts and have stimulated experimental chemists to set to work and discover new chemistry. With the advent of Perturbation and Frontier MO theories, the significance of HOMOs and LUMOs has been extended to account not only for the course of reactions but also for their relative rates. Indeed the orbital symmetry models discussed above can be regarded as subsets of PMO theory.

4.8 Conclusion – hypothetical descriptions of reactions

In this chapter we have outlined many ways in which chemists think about how compounds react together and these individual models have been the predictive basis for new work and the solution of new problems. However there often comes a time when the chemist is faced with a problem that does not seem approachable using any established model. He must then devise a working hypothesis which is both a summary of all that he knows and a prognosis of what seems intuitively right. This is a very frail model unless he is prepared to make explicit what assumptions he has used. It is best to express the whole in the usual language of chemistry so that inter-relationships such as those that we have met in the last two chapters can be perceived. In the remainder of this book we shall encounter examples where the modelling procedures that we have described are applied to a very wide range of chemical situations ranging from pure chemistry to the interfaces of chemistry with biology and with engineering. Let us leave the subject of reactions and their mechanisms with a few words of Konrad Bloch, the Nobel Laureate for Medicine (1964), who assesses the attraction of this field of study for chemists as follows: 'I think it is the very nature of proposed reaction mechanisms that they are romantic' (ref. 51).

MODELLING IN SYNTHESIS, SPECTROSCOPY, AND STRUCTURE DETERMINATION

5.1 Introduction

So far we have discussed the principles of modelling and have described the processes of chemical thought and logic in relation to these principles. The analysis of chemical logic distinguished three types of model which differ from each other in their point of contact with experimental results. It is the recognition of these points of contact and the subsequent application of chemical models which concerns us in the following pages.

The application of models of many types is a great aid to research. We have selected three fields of study for discussion in this chapter: synthesis, spectroscopy and structure determination.

Although synthesis is a problem that chiefly concerns the organic chemist the approaches used have general validity. It is convenient to consider the applications of modelling to synthesis from an organic point of view and to comment upon complementary areas of other branches of chemistry as we go along. However, spectroscopy and structure determination are universal, chemically speaking, and models are applied in the same way whether the substance in question be organic or inorganic. Before we investigate these areas in detail, we would do well to examine what constitutes a viable model in the experimental fields of chemistry just as we did for chemical concepts.

5.2 What constitutes a viable model, and how can we construct it?

In experimental chemistry, the need to construct a model arises as soon as it is no longer possible to predict the outcome of a reaction using all the information available. This information includes the chemist's experience and the chemical literature. If a thorough study of what is known has been carried out it will be easy to complete the essential first stage of constructing a model which is to *define the system*. Since we are always dealing with extrapolations and comparisons it is obligatory to define one arm of the comparison as precisely as possible. We must be clear what our prototype is. Perhaps it is of equal importance to appreciate the limits of certainty of our definition; there is then less danger that we will over-value our results and thereby overstep the bounds of rigour.

However vague we may be about the prototype, we always have the advantage of knowing precisely what our model is. It is obviously pointless working with something which we cannot describe with sufficient exactness, and clarity of thought in defining the system is essential for progress.

A corollary to the need to understand the prototype clearly is the importance of *defining the problem* to be tackled. For example, if we wished to travel between two cities by road but did not know the best route, it would be absurd to choose a map that represented the geology of the terrain between the two cities. This is trivial and common sense, but it makes the point that the model chosen must be appropriate to the problem at hand. Pushing the argument further, to find the distance between the two cities, there is no need to think how we can get from one to the other. In other words, the simplest possible model should be designed; it is economical in effort and much easier to draw useful conclusions if the number of variables is strictly controlled.

The bulk of the examples that follow deal with difficulties encountered in recognising the important variables. It is a matter of judgment, experience and also a little luck. In planning the road journey between two cities, one very significant variable would be the quality of the roads chosen. This is within the control of the driver, but he cannot control others who may wish to use the same route. The traffic density is often difficult to gauge. Similarly if we consider the tetracycline molecule (chapter 3) it is fairly clear from the number of carbonyl groups present that a nucleophilic addition of hydroxide ion, say, is probable. However by inspection it would be very difficult to say what conditions would be required to favour one reaction over possible competitors. In this case we have a practical problem – how to carry out a selective reaction – and we shall return to it.

Identification of variables is the crux of model design whether we are looking at things from a physical, biological or conceptual point of view. We have already discussed how two major theories have had to be modified to include variables at first judged to be unimportant. Collision theory of reactions (chapter 4) neglected the decisive influence that the shape of a molecule has not only upon the rate of a reaction, but also upon its course. It was necessary to abandon the theory because the simple inclusion of an adjustable parameter, the steric factor, offered no better understanding of what was going on. This is not to say that an imperfect model is valueless. The history of the development of theories of electrolytes illustrates well how refinement in experiment leads to refinement in the model. In contrast to the attempts to amend collision theory, the constants that were introduced into Debye–Hückel–Onsager theory were logical embodiments of physical processes that were thought to be occurring when an ion in solution is subjected to an electric field. But as we saw, even these developments were not sufficient to account for all experimental results and as a consequence, the generally useful concept of ion pairing was introduced. We may fairly say that the Debye–Hückel–Onsager theory is a good quantitative model of the conductivity behaviour of ions of low charge in dilute solutions in polar solvents. This statement makes clear what is being taken into account and therefore what is excluded. It combines the three steps necessary in setting up a model. The system is defined as ions dissolved in a polar solvent, the problem is to provide a quantitative explanation of how electrical conductivity depends upon the variables which are solvent, and charge and concentration of ions.

When studying reactions in a synthesis it is just as difficult to select what should be included in a model as it is to provide a theoretical description of a physical phenomenon. The very phenomena that we have just recalled, ion association and dissociation, can profoundly influence the course of a reaction. This is also true of solvent and salt effects, and for this reason, it is important for a synthetic chemist to have a good qualitative understanding of the physical models that describe how molecules behave in solution. Such influences concern the outer environment of the molecule and they interact strongly with the innate properties of the molecule itself, its inner environment; this is a level of complexity that is omitted from physical descriptions of solution properties. It is not surprising therefore, that many chemists are exceedingly sceptical about the use of any model system in experimental work.

In fact, as far as synthesis is concerned, there is only one possible model which precisely mimics its prototype. This model is the enantiomer of the prototype molecule. In all reactions, other than those in which it is placed in an asymmetric environment (particularly important in enzyme catalysed reactions), the enantiomer will behave qualitatively and quantitatively in an identical way to its prototype enantiomer. Because of this, the exact reaction conditions and experimental procedures developed with the model enantiomer will be applicable to the prototype. Clearly it will be rare for this type of model to find profitable use – we do not always have optically active compounds one enantiomer of which is expendable – but one outstanding example exists from the earlier work on Woodward's synthesis of vitamin B_{12} (refer to section 6.7.3). One intermediate containing five chiral centres was obtained in racemic form (figure 5.1) (ref. 1). The synthesis of this intermediate was itself a formidable task requiring the control of stereochemistry and this was possible for four of the five chiral centres. Woodward used the enantiomer which did not correspond to the natural configuration of his target, vitamin B_{12}, as a model and in no case did the behaviour of the prototype differ. In terms of the conceptual model, this is exactly what one would expect since every interaction in the reacting system, in both ground and transition states, must be the same for each enantiomer. The extreme difficulties of modelling a complex system reliably are emphasised by Woodward when he comments that the non-natural enantiomer 'is just about the only kind of model which we regard as wholly reliable'. In view of the problems of identifying and estimating the significance of interactions in a reacting system, many chemists will feel sympathetic with this point of view.

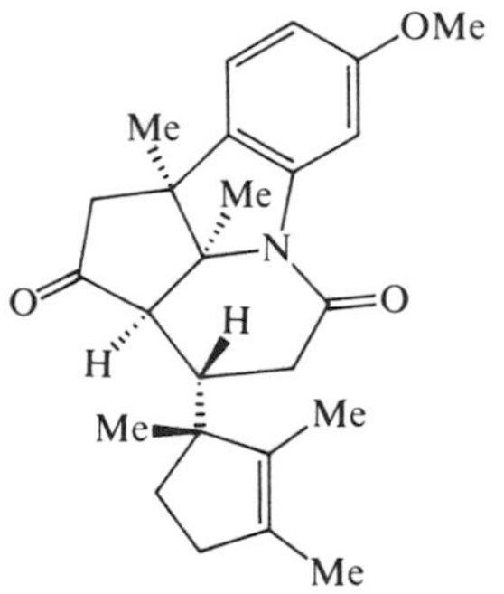

Figure 5.1. An intermediate in Woodward's vitamin B_{12} synthesis.

5.3 Organic synthesis

The ability to synthesise a given molecule is one of the most important accomplishments of an organic chemist and some of the greatest achievements of organic chemistry are to be found in the field of synthesis. The need for synthesis is a feature of all fields of organic chemistry: the synthesis of compounds of biological activity in the pharmaceutical industry, the synthesis of specifically labelled molecules for metabolic and biosynthetic studies, and the synthesis of molecules designed to test mechanistic or theoretical proposals in physical or biological organic chemistry. In addition to these applications, the activity of synthesis itself is a subject worthy of deep study. Synthesis requires the uniting of all the tools available to the chemist in an attack on a specified target and the success of the effort reflects the reliability of the conceptual model and its allies. However, progress in the development of the conceptual model has often been a consequence of synthetic studies, for example, Woodward and Hoffmann's treatment of the concept of orbital symmetry described in the previous chapter was a direct outcome of the early stages of the synthesis of vitamin B_{12}. We shall now analyse the activity of synthesis into stages indicating at each turn when modelling can be of service.

The first stage of any synthesis is obviously the definition of the target molecule according to the overall aims of the project. A synthetic strategy must be developed by dissecting the target molecule into potential simpler precursor molecules, which become intermediate goals. From an evaluation of the possible routes generated by the dissection, the third and practical stage of synthesis can begin, namely the translation of paper drawings into reactions themselves. General discussions of how these stages may be carried out have appeared in the literature (refs. 2, 3) and in this chapter, we restrict our discussion to the impact of modelling on synthesis. The texts cited discuss strategies appropriate to handling the problems of complex stereochemistry, protecting groups and logistics, amongst other important aspects of synthesis.

5.3.1 *The choice of the target molecule*

In company with mountaineers, the chemist's simplest reason for selecting a target is 'because it is there'. Such a target is chosen for the compelling reason that it presents a challenge which will test the science and art of organic chemistry to the full. Vitamin B_{12} is in this class but one could also include molecules which at one time appeared to be pure fantasy like cubane (5.2.1). If this may seem to be a rather arbitrary method of project selection, it should be realised that projects of such origin have led to some of the most important advances of chemistry as a whole. The challenge of synthesising a complex natural product or an unusual molecule that exists only on the drawing board will always remain an important activity for organic chemists, an activity comparable with the setting of new records in athletics or the scaling of hitherto unclimbed peaks. There is always scope for creativity and work of elegance in organic synthesis; the subtle balancing of interactions between different atoms, molecules and functional groups is an art which has its own classics, epics and folk-lore.

Alternatively, a target molecule may be required for a particular purpose. A physical organic chemist may wish to synthesise a molecule designed to exhibit a specific pattern of behaviour. For example the unsaturated tosylate (5.2.2) was synthesised to examine the effects of π-orbital participation in solvolysis reactions and it was compared with the reference substance (5.2.3), a saturated tosylate, considered as being incapable of exhibiting such effects (ref. 4) but having a similar, though not identical geometry. Such a synthesis may represent no

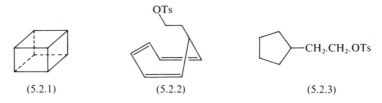

(5.2.1)   (5.2.2)   (5.2.3)

Figure 5.2. Synthetic targets.

difficulties in practice but in the choice of the target, all relevant important elements in the system must be taken into account if the results are to be meaningful. Further examples of this type of work were considered in connection with effects in chapter 3.

Another example where the conceptual model defines the target for synthesis is in the chemistry of the annulenes (ref. 5). According to the Hückel molecular orbital theory, a cyclic, planar, conjugated molecule containing $(4n+2)$ π-electrons should have aromatic character. A system containing 18 π-electrons for example satisfies the Hückel criterion for aromaticity and the synthesis of such a cyclic molecule presents a challenge to the synthetic chemist (5.3.1). Sondheimer and others have synthesised many $(4n+2)$ π-electron systems, recent examples being the unusual anion (5.3.2) and even analogues of heterocyclic aromatic compounds (5.3.3) (refs. 6, 7). These molecules require more elaborate syntheses than the molecules discussed in the previous paragraph.

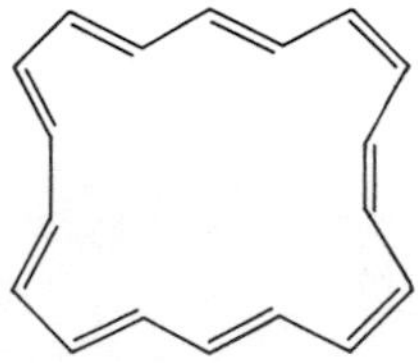

OMe

H
N

(5.3.1) 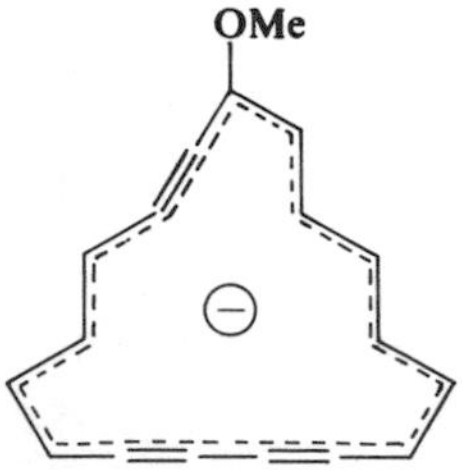   (5.3.2) 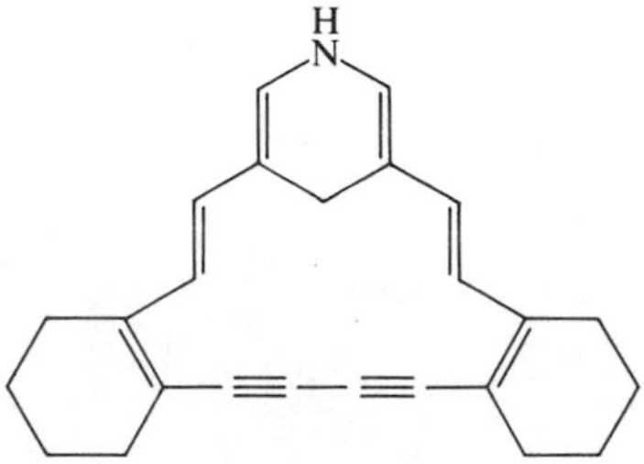   (5.3.3)

Figure 5.3. Annulenes.

Other theoretical considerations may lead to the choice of a target for synthesis. For example, it may appear that a particular compound could be involved as an intermediate in a biosynthetic sequence and that it may be present, albeit in minute amounts, in nature. An isotopically labelled form of this intermediate then becomes the target for synthesis so that it can be fed to the organism in question (see chapter 6) and its biological fate determined. A case in which mechanistic considerations highlighted the possible involvement of an intermediate concerns the biosynthesis of colchicine (ref. 8). The *in vitro* analogy of the ring expansion of the diene (5.4.1) to the seven membered ring of compound (5.4.2) suggested that a parallel ring expansion of the newly isolated alkaloid androcymbine (5.4.3) as its O-methyl derivative to give colchicine (5.4.4) takes place in nature. Thus it was necessary to synthesise O-methylandrocymbine in labelled form. In this case, the synthesis is trivial involving O-methylation of naturally produced androcymbine with a methylating agent bearing a tritium label. However, it is more common for syntheses of specifically labelled precursors designed to test biosynthetic hypotheses to be lengthy and complex.

A compound may also become the target of synthesis in order to produce a specific range of effects; here effect is used in the sense of Baines *et al.* (ref. 9). The design of a molecule to have required biological action may be based on the supposed mechanism of its interaction with an organism. This may be the case if the molecule is expected to have an effect upon a particular enzyme the properties of which are known, and this approach is rapidly developing, but it has been much more usual to adopt as model some crude form of structure–activity correlation derived largely empirically (refs. 10, 11). Such a correlation is an attempt to extend the analysis of a molecule's

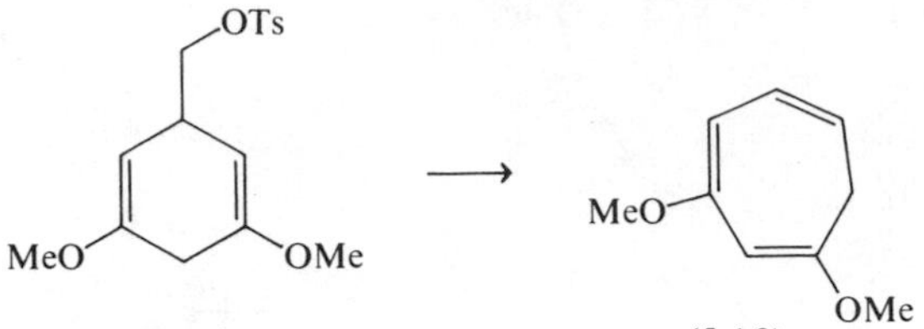

(5.4.1) (5.4.2)

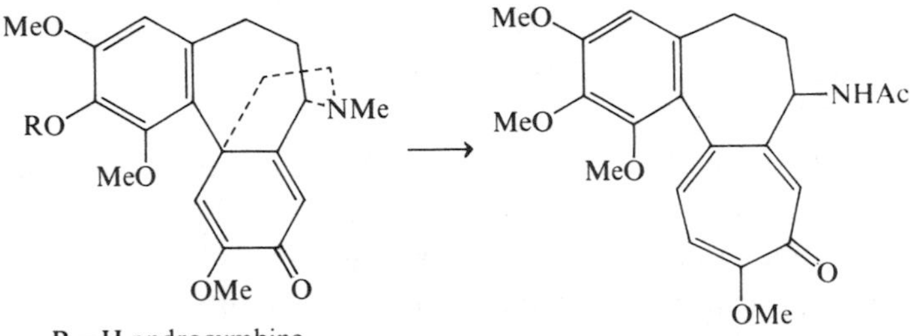

R = H androcymbine
= Me O-methylandrocymbine

Colchicine

(5.4.3) (5.4.4)

Figure 5.4. The biosynthesis of colchicine and models.

structure based upon the conceptual model to include a correlation with its biological activity and hence to incorporate biological action into the range of properties which the conceptual model is capable of predicting. Clearly these models can only be very approximate because the full complexity of the mechanism of interaction of the molecule with the biological system is unknown. This often leads into a broadening of the synthetic target to encompass a range of molecules, variations on a promising theme, which are then analysed by screening procedures. A screen is set up to identify compounds with the required properties and in this way, it acts as a model of the desired end result. A more detailed discussion of this aspect of modelling will be found in chapter 8 (ref. 9*a*).

5.3.2 *Synthetic planning and strategy*

The growing interest in the systematisation of organic synthesis into forms which may be capable of computerised handling implies a high state of organisation of the present day organic chemists' conceptual model (refs. 2*a*, 12). There are usually more possible routes to a complex molecule than a chemist is likely to spot by the time-honoured method of sitting down with a pencil and paper, essential though this is. Be the approach individual or formalised there are common points in the application of modelling and the general procedure is as follows. The chemist breaks down the target structure into potential precursor molecules that can be combined by known methods or by plausible novel reactions into the target molecule. He continues the dissection and recombination until the precursor fragments are readily available molecules from commercial or other sources. The chemist's problem is not only to generate these series of precursors but also to choose the route most likely to succeed. It then remains to join in the laboratory the fragments so generated. Experience in the execution of the plan in the laboratory may require that modification to the original plan be made and the best strategies embody an element of flexibility. We shall now examine the events outlined above in greater detail.

5.3.3 *The dissection of the target molecule*

Before the advent of deliberate systematisation of synthesis, the blend of logic that was used to plan a synthesis represented an exercise which we have called the application of the conceptual model. At the dissection stage, it is concerned with the selecting of building blocks capable of being cemented together into the required target. In many cases it is obvious even to the inexperienced eye what these building blocks should be. For example, if one wished to synthesise the porphyrin (5.5.1), one would instinctively think of pyrroles (5.5.2) as suitable starting materials because they are accessible, relatively stable yet reactive sub-units of a porphyrin: they have the *quasi*-independence that is characteristic of all useful sub-systems. Pyrroles are in fact the basis of all porphyrin syntheses (ref. 13). Other dissections, less obvious, imply the use of a particular reaction type to set up a vital structural detail of the molecule. Typical of this is the addition of a carbene to a double bond to form a cyclopropane ring: this has been applied both intermolecularly in a synthesis of presqualene (5.5.3) (ref. 14) and intramolecularly in a synthesis of sesquicarane (5.5.4) (ref. 15). Notice here that although the same reactive intermediate, a carbene, is involved, it is produced by different methods in each case and this indicates the existence of the desired element of flexibility possible in these syntheses.

Apart from his own intuition and foresight, a chemist may gain hints on how to approach a synthesis of a natural product from the mode of biosynthesis of the target molecule. A synthesis patterned after the

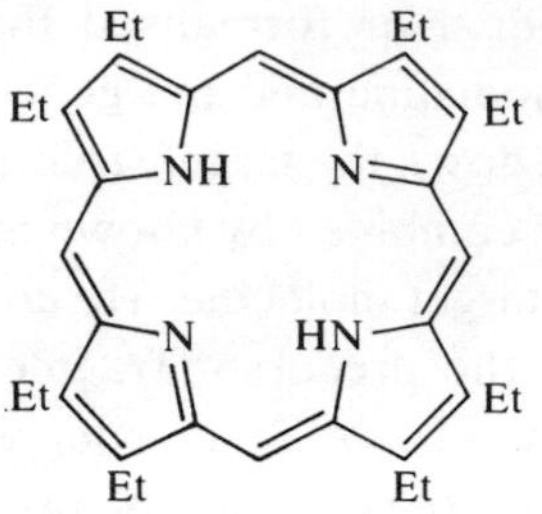

(5.5.1)

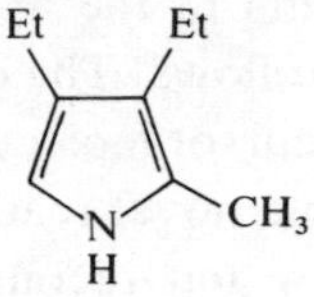

(5.5.2)

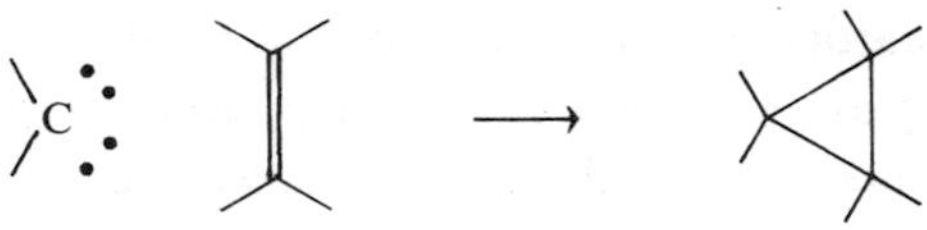

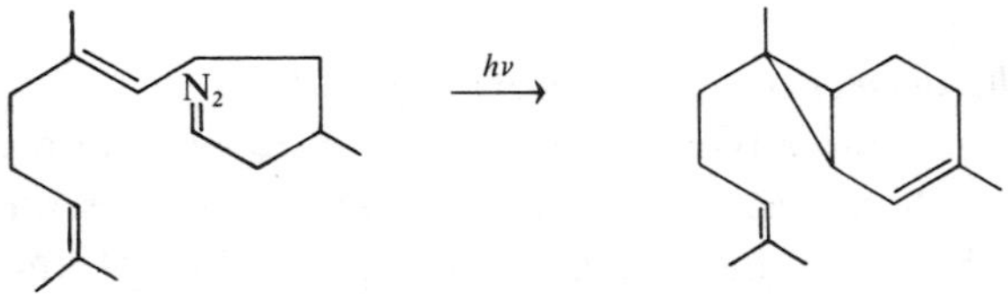

Sesquicarane

(5.5.4)

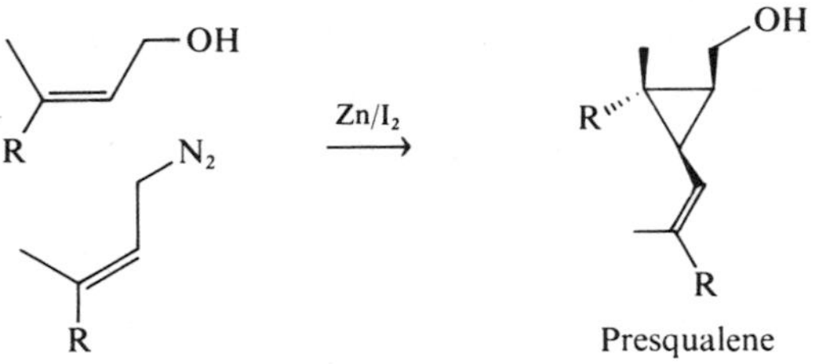

Presqualene

(5.5.3)

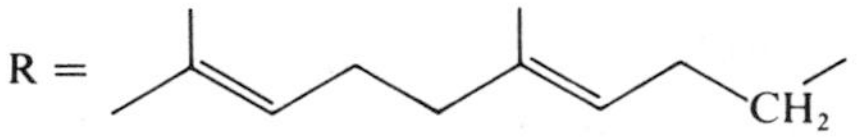

$$R =$$

Figure 5.5. The dissections of some syntheses.

route that nature chose is a model of the biosynthesis and this biogenetic approach to synthesis will be discussed in chapter 6 with reference to examples from the steroid and alkaloid fields.

A useful source of molecular history which can be a stimulus to the design of a synthesis is the reactions of the target molecule on degradation. The degradation at the very least shows what conditions not to use in the final stage of the synthesis but can more usefully indicate the stability of various fragments which could be recombined. For instance the early attempts to synthesise the $4n$ π-electron cyclic diene cyclobutadiene (5.6.1) were carried out under conditions too drastic for the product to survive. The *cis*-dichlorocyclobutene (5.6.2) is an obvious attractive precursor (ref. 16) but on treatment with metal amalgams, only dimers of cyclobutadiene can be detected (5.6.3). In fact cyclobutadiene has only been detected under the mildest of conditions: oxidation of the iron tricarbonyl derivative (5.6.4) with ceric ion yields free cyclobutadiene as shown by subsequent trapping experiments (ref. 17) and photolysis of the pyrone (5.6.5) in an argon matrix at very low temperature leads to cyclobutadiene which can be characterised by spectroscopy (ref. 18).

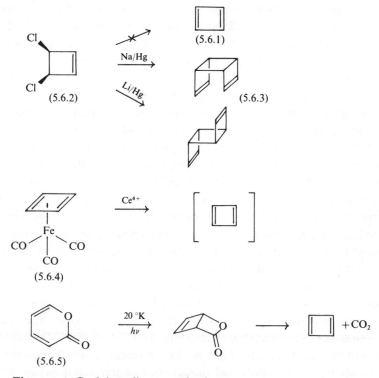

Figure 5.6. Cyclobutadiene synthesis.

Computer-assisted design of synthesis. A very expensive assistance to the chemist in designing synthesis is being studied by Corey (ref. 19) who is developing a set of algorithms for each reaction type for use in a computer-assisted dissection of the molecule. The virtue of computer planning is that the computer will suggest dissections that the chemist may have overlooked or instinctively rejected as unworkable. The computer-aided synthesis works like this. The chemist tells the computer the target structure and the computer then methodically breaks the target down into fragments by exhaustively cleaving one or two bonds at a time. The potential of the generated fragments as precursors in a synthesis is assessed by the machine according to weightings for the various types of dissection which have been built into the programme, that is by evaluating the viability of the reverse of the dissection. In this way, the experience of the chemist is built into the programme and the computer is guided by the common knowledge that, for example, the Diels–Alder reaction is very reliable for constructing bicyclic systems. A danger inherent in this approach is that the formalisation of existing patterns of thought may suppress new ones, but if this warning is heeded the exhaustive analysis of which the computer is capable more than compensates for any loss in originality. The computer then repeats the process on the intermediate it generated building up a tree of syntheses, branches of which terminate when recognisable fragments or known compounds appear as the sub-goals.

It has been observed in industrial plant in which process optimisation is being assisted by computer programmes that the plant operators learn from the computer. They build the pattern of plant response to operating conditions that is suggested by the computer into their own private models of plant behaviour. It is likely that the chemists' models will, in similar manner, be enriched by studying ideas generated by the computer and that this, in turn, will lead to the writing of more effective and more comprehensive computer programmes.

In both Corey's and Henrickson's related work (ref. 12), abstraction continues past the level of structural formulae into a realm presently beyond the reach of chemical intuition. They are stretching the conceptual model to its abstract limits just as the synthesis in the laboratory stretches it to its practical limits. There remain nevertheless points of contact with the conceptual model, one in particular being the need for labels for reactions in whatever formalism used. Corey in fact names his algorithms after the parent reactions. Chemistry has grown up with reactions bearing the names of their discoverers; most such reactions are very well known and are readily embodied into the conceptual model. A name comes more readily to the human mind than an abstract drawing of a structure and hence the named reaction serves as a link in the synthetic chain between developing a plan and the reconstruction of the target molecule following it.

5.3.4 *Named reactions*

The omnipresence of the named reaction shows once again the extent to which the organic chemist uses analogy and precedent in forming his synthetic plans. A named reaction is usually one of proven synthetic applicability, that is it has been used successfully in a wide variety of examples with good yields. Snags in the form of competitive side reactions will frequently be known and the procedure can be adopted with confidence. But not all named reactions are of synthetic use; some names have been added by the discoverers or by close admirers in the hope of immortality and others are applied to reactions of very limited scope. The most useful named reactions are exhaustively discussed in the series of books entitled *Organic Reactions* (ref. 20).

From the point of view of synthetic strategy, named reactions fall into two classes: The first class allows a very wide range of carbon skeletons to be built up, for example the Diels–Alder and Michael reactions. The former in particular has many attributes including the absence of side reactions and, most importantly, the predictability of the stereochemistry of the reaction. Control of stereochemistry, as the reader will have gathered from the reference to vitamin B_{12}, is a corner stone of organic synthesis and a common method of holding two centres together in the required stereochemical relationship is to join them in a ring. The Diels–Alder reaction achieves this through the mechanism of an electrocyclic process which is now well understood (chapter 4) and a recent application is Corey's range of syntheses of the prostaglandin hormones (5.7.3) (ref. 21). He sets up the relative stereochemistry of the centres marked a, b, and c in this one reaction between the diene (5.7.1) and the activated olefin (5.7.2), although many further steps are required to complete the synthesis.

Griseofulvin synthesis. As a further illustration of the use of a named reaction as the basis of a synthesis this time involving the very resourceful Michael reaction, Stork's synthesis of the antibiotic griseofulvin (5.8.4) (ref. 22) shows a characteristic concise elegance in the

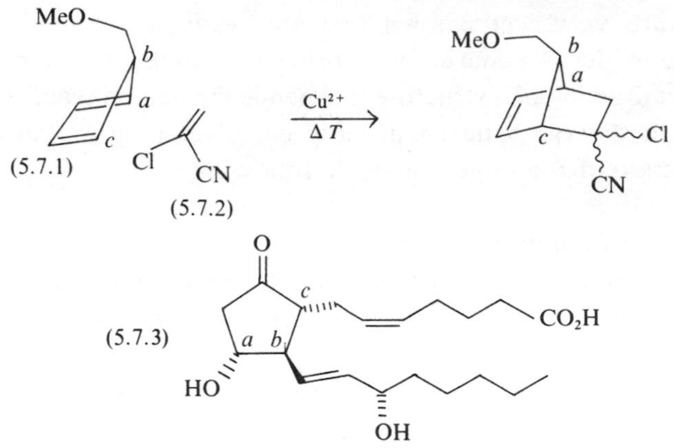

Figure 5.7. The synthesis of prostaglandins.

directness of the approach. The coumaranone (5.8.1), which can form an enolate anion in basic conditions, was condensed with the doubly unsaturated ketone (5.8.2) in a Michael reaction yielding the intermediate (5.8.3). This product can again undergo a Michael condensation, this time intramolecularly to form griseofulvin (racemic) directly. There is the impression of a hunch that paid off about this synthetic plan when the many potential side-reactions are considered. It is obviously possible for the first Michael addition to occur into the olefin rather than, as was desired and found, into the acetylene. Not only this, but the stereospecificity of the second Michael addition is surprising since the two possible C-2 epimers of (5.8.4) have *a priori* roughly the same stability. Stork pointed out that the conformation of the intermediate primary Michael adduct (5.8.3) which will react more rapidly under kinetic control of the reaction will be the one in which the overlap of the enolate anion (donor) and the π-system of the unsaturated ketone (acceptor) is largest, and hence the transition state for reaction in this sense lowest. Fortunately, this corresponds to the conformation giving rise to griseofulvin.

Synthesis of structural classes of compounds. The second class of reactions that have been accorded names are of more limited use in that they are intended for the synthesis of a particular class of compounds. This is especially the case with heterocyclic organic compounds. If faced with the synthesis of an alkaloid based upon an isoquinoline structure (for example norlaudanosine, chapter 6, figure 6.2), the Bischler–Naperialski synthesis would immediately come to

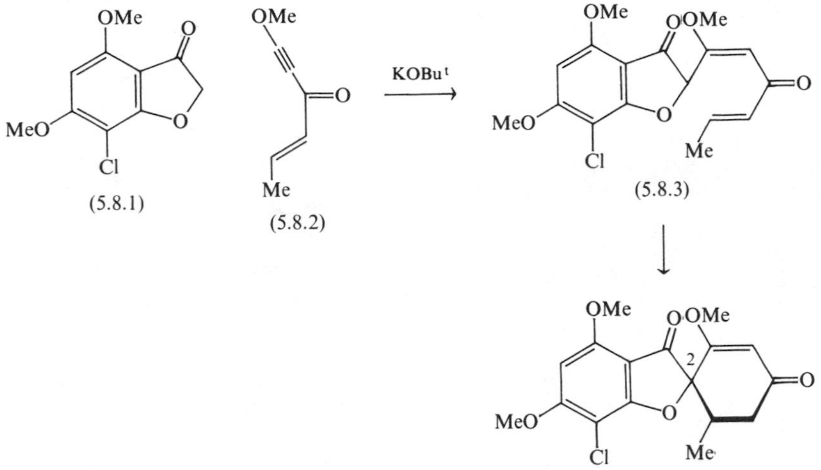

Figure 5.8. The synthesis of griseofulvin.

mind (figure 5.9) (ref. 23) or for the porphyrin synthesis mentioned earlier, Knorr's pyrrole synthesis would certainly be considered (figure 5.9) (ref. 13).

The success of a synthetic method can lead it to become a named reaction in a very short time and the classic example of the last twenty years is the Wittig olefin synthesis (ref. 24). In its original form, the Wittig reaction was limited to the synthesis of 1,1- or *cis*-1,2-substituted olefins, but recently, several groups of researchers have been able to develop the successful reaction type into more flexible syntheses so that now it is possible to prepare amost any configuration of di- and tri-substituted olefins using the same basic reaction (refs. 25, 26). Figure 5.10 illustrates the possibilities, which have been extensively applied to natural product syntheses.

In terms of our earlier analysis of the properties of an organic molecule or of a reacting system, why are synthetic methods such as named reactions so successful? Clearly in these cases the behaviour of the reagents can be predicted accurately by analogy. This means that in all these reactions the significant interactions between reagents, substrates and environment are the same. Named reactions can therefore give us valuable clues as to what the significant interactions are and from this position, new useful chemistry can be developed. However, the reliable reactions do not always bring about the required

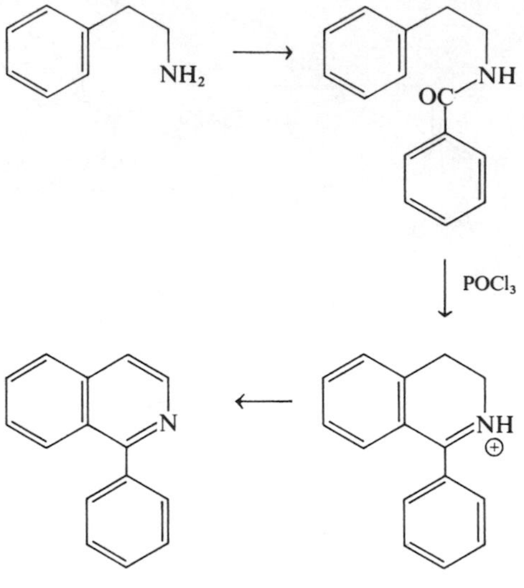

Bischler–Napieralski isoquinoline synthesis

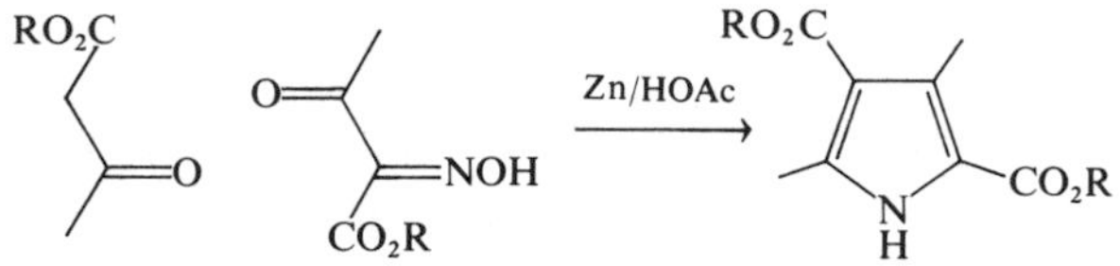

Knorr's pyrrole synthesis

Figure 5.9. Syntheses of heterocyclic compounds.

transformation and it can happen that a molecule does not lend itself to synthesis along one of these well-trodden paths. In such cases, the way is open for the introduction of new synthetic techniques developed with a particular target or group of targets in mind.

5.3.5 *New synthetic reactions*

If a direct analogy cannot be found in the literature then the predictive power of the conceptual model must be brought into play to design new synthetic reactions. In the early 1960s such a position was reached with olefin synthesis because it was found that methods were just not available for the synthesis of many types of terpenoid, especially acyclic compounds. However subsequent research including extensions to the Wittig reaction has completely transformed the

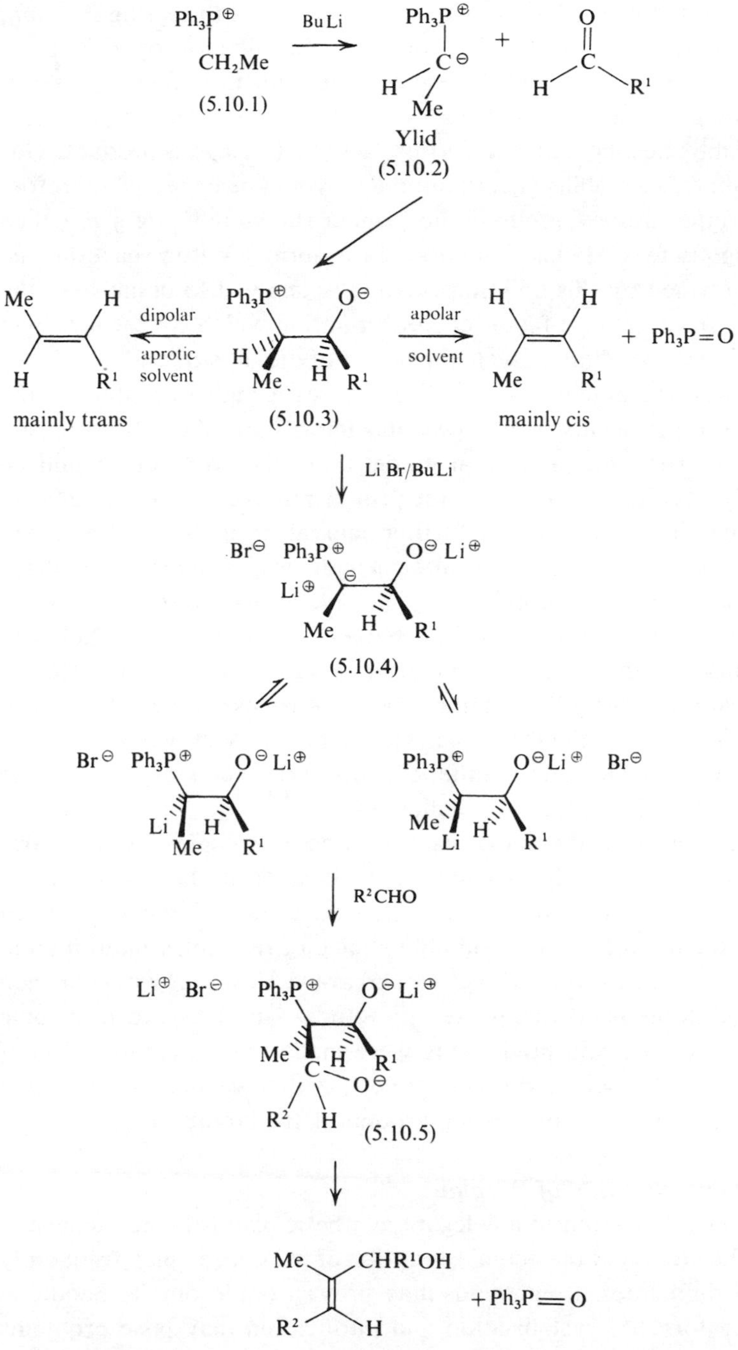

Figure 5.10. The Wittig olefin synthesis and modifications.

situation (ref. 25). The logic of Corey's extension of the Wittig reaction, for example, shows the simple principles of analogy in operation.

Wittig showed that the treatment of a tetrasubstituted phosphonium salt bearing an α-hydrogen atom (5.10.1) with base causes deprotonation of the phosphonium salt forming an ylid (5.10.2) as product. This intermediate is a stabilised carbanion and would be expected therefore to add to the carbonyl group in the manner shown in figure 5.10 which is analogous to a Michael addition. In a normal Wittig reaction, the betaine formed by this addition (5.10.3) is allowed to collapse to the products from its most favoured conformation which is very sensitive to the reaction conditions and to the nature of the substituents, R. Corey noticed that the betaine itself still possesses a hydrogen atom at the carbon α to the positively charged phosphorus and there is no reason to believe from the conceptual model that this hydrogen should be markedly less acidic than the first proton removed. Accordingly, he treated betaine (5.10.3) with a further equivalent of base and obtained the β-oxidoylid (5.10.4). This intermediate is yet another stabilised carbanion that could conceivably be added to another aldehyde, R^2CHO, yielding a third and final betaine (5.10.5) which is allowed to collapse to the desired product, a trisubstituted allylic alcohol. Under carefully controlled conditions, the base used, the solvent, and at low temperature, the reaction is stereospecific. Corey is a specialist both in developing new synthetic procedures such as this and in demonstrating their practical application.

We have seen in the above examples how analogies from various sources can influence the design of a synthesis from the moment of its conception as a worthwhile target. Several times reference has been made to the importance of a suitable choice of reaction conditions and to other respects in which things can go awry. However we have been interested in the plans themselves up to now, and let us turn to some cases where confident predictions were made from a combination of the conceptual model and other models, which we shall discuss, and see how they fared in the testing ground of the laboratory.

5.3.6 *The execution of the plan*

The deduction of an elegant synthetic plan is by no means the end of the story and the actual execution of the chosen plan frequently presents difficulties. Compounds may prove troublesome to handle in the laboratory or crystallisation and purification may pose problems or an intermediate may be unstable under normal conditions. These practical difficulties often cannot be predicted easily by the conceptual model and syntheses have to be modified to overcome problems of handling intermediates. One of the fascinations of organic synthesis is that intermediates in a synthesis are very often new compounds which introduce the element of surprise because their exact behaviour under specific conditions will not be known. It is relatively easy to predict the type of reactivity to be expected but, as was pointed out in section 5.2, the prediction of the degree of reactivity requires knowledge of a close analogy that defines reaction conditions rigorously as an inspired hunch. For this reason, it is often necessary to test proposed steps in a synthesis before any attempt is made to carry out the entire concept. Such a situation arises when a proposed step is new and no good practical analogy can be found for it. Usually it is impossible for reasons of availability or expense to test the critical steps of a synthesis on the compound for which these steps are intended. In these cases it becomes necessary to turn to model compounds. Model compounds used in this sense reflect a failing of the conceptual model to give an adequate prediction of the ease of a reaction, or they indicate the absence of any satisfactory precedent. In this context we meet the use of the term 'model' in one of its commonest connotations in chemistry.

Before embarking on a study of the critical steps it is always as well to be certain that no previous evidence recorded in the literature can help the study further. Such evidence may provide 'off-the-peg' models which require little further experimental work save perhaps a few minor alterations. The analysis of the literature model follows the same pattern as the following description of model compounds which are tested in the laboratory.

A chemist's own earlier work may be a source of a close analogy for a crucial step and frequently one finds syntheses designed around reactions which the chemist himself has found to be reliable. This is one origin of style in synthesis found in the work of a particular chemist or school.

Lycopodine. An example of this type of 'off-the-peg' model was applied to the synthesis of the alkaloid lycopodine by Stork (ref. 27). The strategy required the cyclisation of the nitrogen heterocyclic compound (5.11.1) to lycopodine (5.11.2), under acid catalysis in a stereospecific manner. As a model, the carbocyclic compound (5.11.3) was chosen which was known to cyclise under acid catalysis in the sense shown to form the bridged compound (5.11.4). The mechanism can be expected (conceptual model) to involve the generation of unsaturation (5.11.3) by enolisation followed by attack on the aromatic ring in an intramolecular electrophilic substitution reaction yielding the product. However when the same conditions were applied to the prototype nitrogen heterocycle (5.11.1), the products obtained did not correspond to the expectations based upon the model (5.11.5, 5.11.6). The reasons for this deviation can readily be suggested by considering where the model chosen may not have satisfied our stringent criteria of viability (section 5.2). Firstly, what are the differences between the prototype and the model? One difference is the O-methyl substituent in the aromatic ring of the prototype but we would not expect this to hinder the desired reaction, indeed it activates the positions *ortho* and *para* to it to electrophilic attack which presumably occurs in the model reaction. Another significant difference is the nitrogen substitution of the prototype, which, being α to the expected enol intermediate, could conceivably open alternative reaction pathways, as the operation of the reaction showed. It is stimulating to apply the conceptual model to rationalise what went wrong and why the undesired products were formed. Fortunately, it was possible to modify the prototype (5.11.1) to the derivative (5.11.7) which underwent the required cyclisation.

O-Methylandrocymbine. When no analogy is available from the literature or from earlier work, or when such analogy as exists is inconclusive, the testing of the feasibility of a proposed reaction must be carried out upon a model compound. It was pointed out in section 5.2 that only one very specific model compound, the enantiomer of the prototype, can be completely reliable, but provided one takes good account of the transforms between model and prototype, a model compound can provide valuable information. The choice of model compound is often one of convenience, a fact which unfortunately shows neglect of a thorough consideration of the prototype–model transforms. The earlier example of the synthesis of O-methylandrocymbine offers a hypothetical example (figure 5.4). In order to establish the optimum conditions for the methylation of the phenol, one would need to carry out trial runs, but androcymbine is only available in small quantities whether from natural sources or from synthesis. It would clearly be an extravagance to expend valuable androcymbine in development stages. In any case, the chemist would be wise to carry out a 'cold run' on the prototype

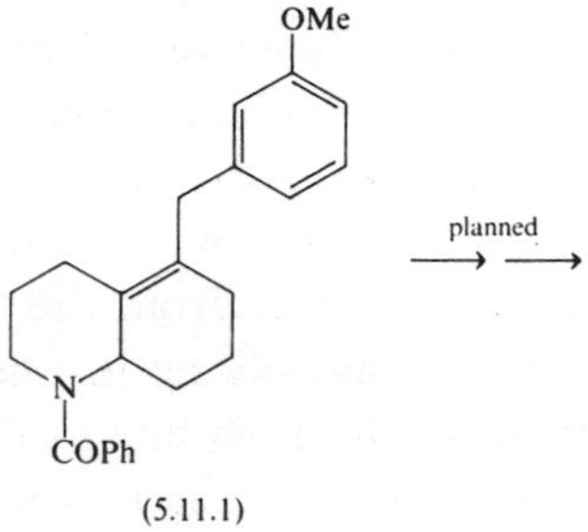

(5.11.1)

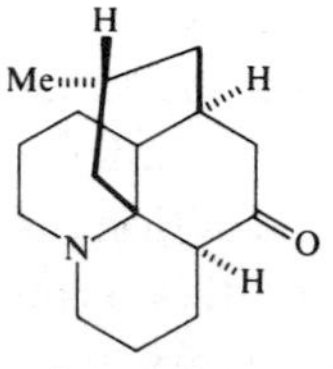

Lycopodine
(5.11.2)

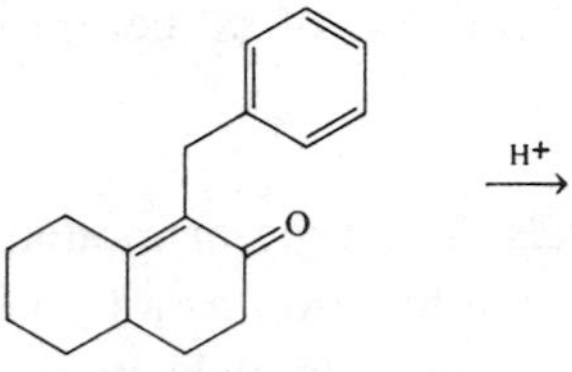

Model system
(5.11.3)

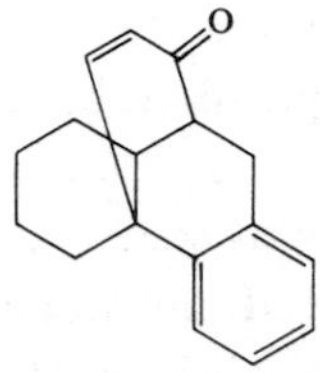

(5.11.4)

(5.11.1)

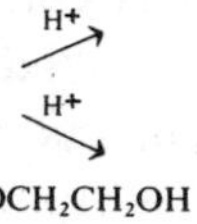

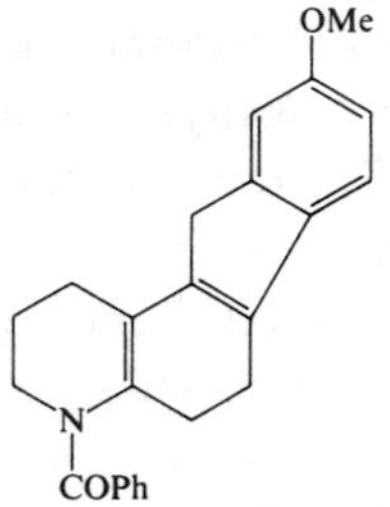

(5.11.5)

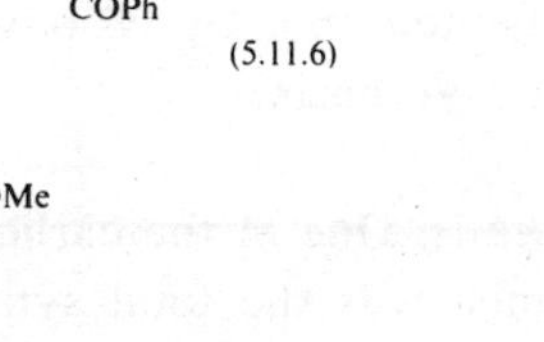

HOCH₂CH₂OH

(5.11.6)

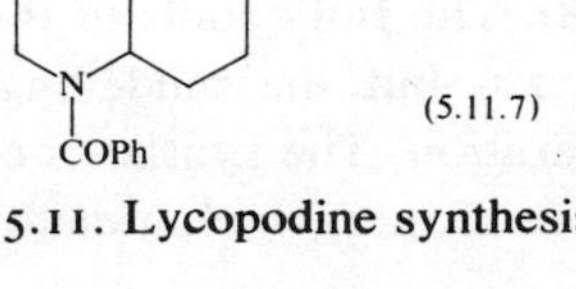

(5.11.7)

Figure 5.11. Lycopodine synthesis.

itself before attempting to synthesise a radioactively labelled O-methylandrocymbine as was required. This common procedure in radiochemical synthesis assumes that the inactive run is a perfect model of the active run. Since the concentration of radioisotope is usually very low and reactions occur mainly away from the labelled site during a synthesis, this assumption can usually be justified. But caution is always necessary when there is the possibility of kinetic isotope effects operating or when asymmetric labelling is being used. In the latter case, an inactive model can give no indication of the effect of a particular step on the stereochemistry of a centre that is asymmetrically isotopically labelled.

Off-the-peg and made-to-measure models. The type of modelling, for which we have selected androcymbine methylation as a hypothetical example, falls into the class of 'off-the-peg' models in which the difference between prototype and model can be wide but is not expected to be significant and allows the chemist to choose the most convenient. Such a choice is not always possible or desirable and it is then better to design a model for the particular purpose to hand. In these 'made-to-measure' models, the chemist is in full control of all the transforms and hopefully can avoid some of the problems caused by the un-avoidable lack of correspondence between the interactions within the prototype and the model which are found with 'off-the-peg' models be they from the literature, the chemist's own work, or readily available sources.

One problem about discussing any kind of model in synthesis, once the application of the conceptual model in the planning stage has been covered, is that modelling in subsequent stages is frequently not published because it is a preliminary study. However the following examples of 'made-to-measure' models should illustrate that model compounds can be useful and do not always produce the unpredictable results commented upon by Woodward and illustrated by the example of lycopodine synthesis.

Steroid synthesis. One of the earlier major successes of the synthetic organic chemist was the total synthesis of the steroid skeleton, for example estrone (5.12.1) (ref. 28). The full details of these syntheses are discussed by Ireland (ref. 2b) but the modelling used in the analysis deserves further consideration. The synthesis of estrone was planned via the intermediate (5.12.2) into which a methyl substituent must be added at C-13 in a stereospecific manner. W. S. Johnson chose as a model for these last stages, when the intermediate itself would be a precious commodity, the readily available decalone (5.12.3) which can be prepared in three stages from α-naphthol. This compound has all the essential characteristics of the advanced intermediate (5.12.2), namely the rigidly fused six-membered rings with a keto group α to the ring junction. Thus C-9 of the model corresponds to C-13 of the prototype and it is here that we must methylate. It is not difficult to recognise why this model proved to be very successful. One would not expect rings *A* and *B* of (5.12.2) to exert any influence on C-13 since firstly, they are devoid of functionality which could give rise to competing reactions, and secondly, the flat fusion of the six-membered rings of the steroid skeleton makes steric hindrance of methylation by these appended rings impossible.

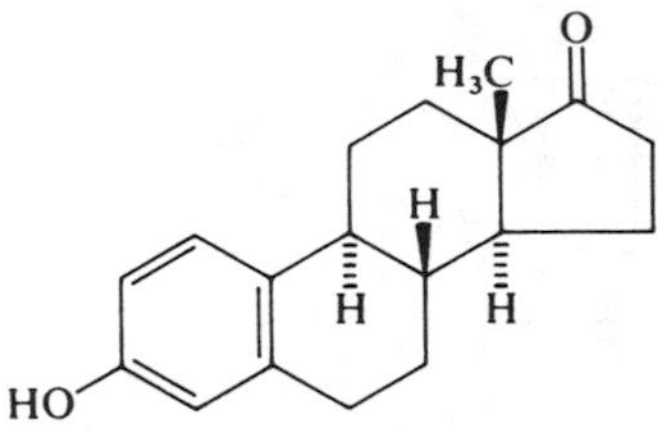

(5.12.1)

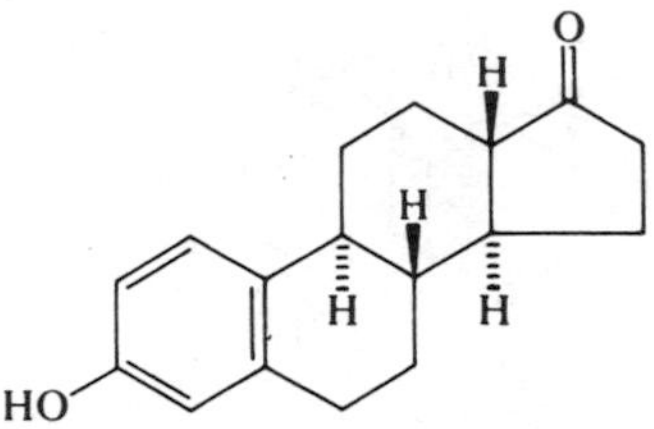

(5.12.2)

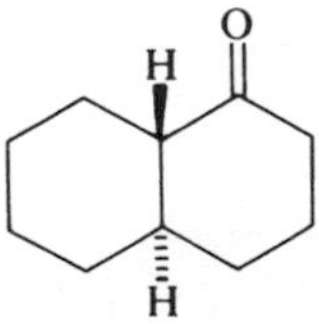

(5.12.3)

Figure 5.12. Steroid synthesis.

Annotinine. The feature of the alkaloid annotinine (5.13.1) which presents the greatest synthetic challenge is the four-membered ring which Wiesner proposed to introduce by the generally applicable reaction of the photochemical addition of two ethylene derivatives (ref. 29). A similar approach was also used by Corey for the introduction of a four-membered ring in his synthesis of caryophyllene (5.13.2) (ref. 30). Wiesner proposed to carry out this photochemical reaction on the intermediate (5.13.3) but this proved to be difficult to synthesise and, in order to test the critical step, he selected the model (5.13.4) which differed only in the absence of the third six-membered ring. On reaction of (5.13.4) with propylene under irradiation, the products obtained were the isomers (5.13.5) and (5.13.6). The ratio of the isomers formed could be altered by using the *N*-substituted counterpart of the model (5.13.4) but in no case could the reaction be controlled to give only the desired direction of addition which corresponds to the model product (5.13.5). Thus the model clearly demonstrated the feasibility of the desired step and also alerted Wiesner to the problems which could be expected from side reactions. Nevertheless Wiesner went ahead and his boldness was rewarded because when the intermediate (5.13.3) was treated with propylene under the same conditions, only one product was obtained, the desired isomer (5.13.7). The frequent use of the photochemical dimerisation of olefins to form four-membered rings has the

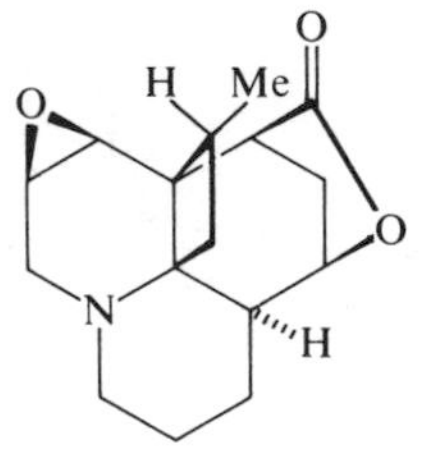

Annotinine

(5.13.1)

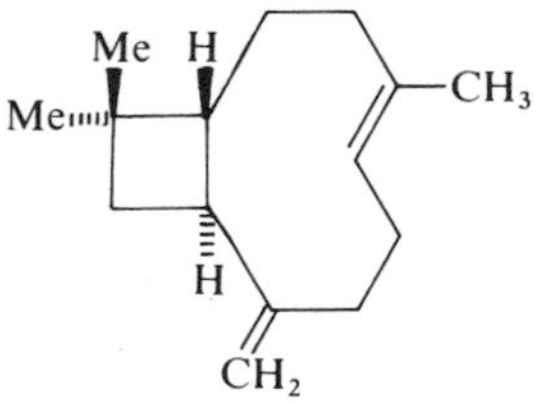

Caryophyllene

(5.13.2)

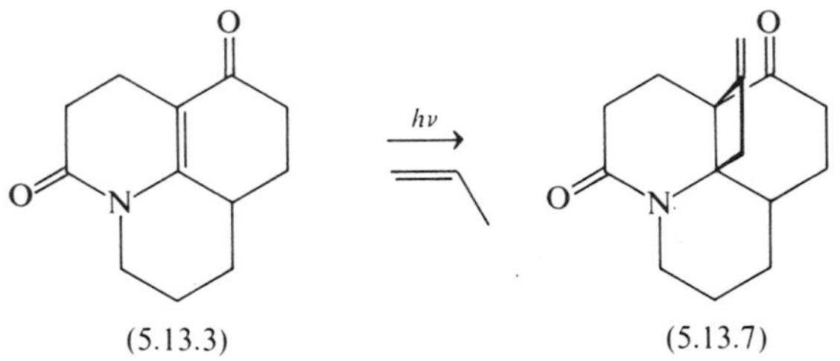

(5.13.3)

(5.13.7)

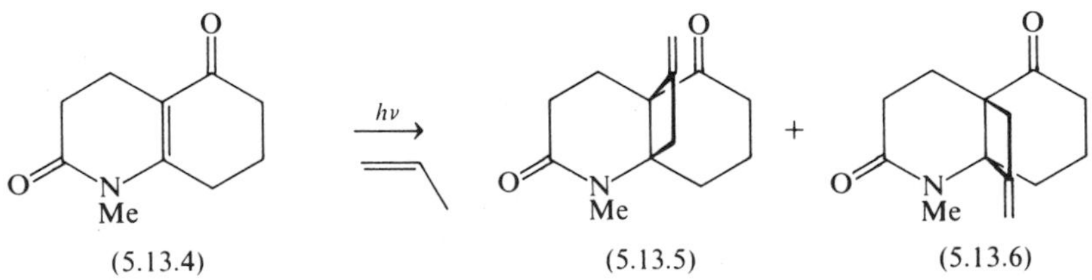

(5.13.4)

(5.13.5)

(5.13.6)

Figure 5.13. Annotinine synthesis.

advantages common to all electrocyclic reactions, as we saw with the Diels–Alder reaction. Such reactions are often reproducible in widely differing molecules without need for elaborate arrays of protecting groups and a wider scope for the choice of a model compound is available.

Vitamin B$_{12}$. The very size and complexity of a molecule such as vitamin B$_{12}$ dictates that advanced intermediates will only be available in very small quantities and that they will require much labour in synthesis, and therefore, modelling will be essential for many of the steps. We have already referred to one reliable model used by Woodward's group in their early work on this synthesis (section 5.2) and we shall now briefly discuss one further example, but many more cases are described by Woodward in his periodic reports to IUPAC meetings on the progress of his synthesis (ref. 31). The results of these carefully planned model studies have led Woodward to write: 'Time and again in the course of studies on the synthesis of vitamin B$_{12}$ we have found that models are something less than realistic – often strikingly so.' As Woodward has warned us, a model selected purely on the grounds of convenience without a full consideration of the factors involved in the prototype–model transforms will be a fruitless exercise more often than not.

Our last example from the synthesis of vitamin B$_{12}$ concerns the hydrolysis of the lactone ring of α-corrnorsterone (5.14.1) and the model used was of the 'off-the-peg' type. It was known that the pentacyclenone (5.14.2) could readily be cleaved, but application of the same conditions which cause the cleavage of the lactam in (5.14.2) to the prototype (5.14.1) had absolutely no effect. The reason for the inertness of (5.14.1) was thought to be the unfavourable steric interaction of the acetate and propionate side chains in the desired product (5.14.3). These interactions are absent in the product of hydrolysis of the model compound (5.14.2). Interestingly, it was found that this unfavourable interaction in the ring opened product was also absent in the epimer (5.14.4), β-corrnorsterone which was readily hydrolysed under the same conditions as the model (5.14.2) is hydrolysed.

We have drawn much from the epic synthesis of vitamin B$_{12}$ and it should not go without saying that the collaboration of Woodward in Harvard and Eschenmoser in Zurich has led to the triumphant completion of two syntheses of this complex molecule (ref. 32).

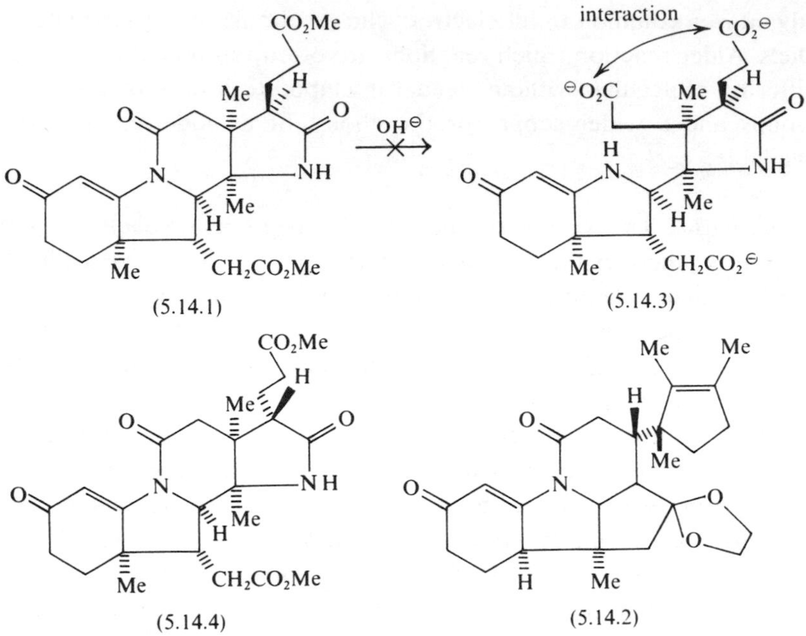

CO₂Me / Me / H / O / O / OH⊖ / N / NH / Me / H / Me / CH₂CO₂Me

(5.14.1)

interaction → CO₂⊖ / ⊖O₂C / Me / H / O / H / N / NH / O / Me / H / Me / CH₂CO₂⊖

(5.14.3)

CO₂Me / H / Me / O / O / N / NH / Me / H / Me / CH₂CO₂Me

(5.14.4)

Me Me / H / O / N / Me / O / O / O / H / Me

(5.14.2)

Figure 5.14. α-corrnorsterone.

5.4 Molecular models

In the example of the vitamin B_{12} synthesis just given, examination of a molecular model would clearly show great differences in the *proximity* of the acetate and propionate groups, but it would be very difficult to decide just from the model that the difference would inhibit the required reaction completely. This underlines the problems associated with both molecular models and structural formulae when the chemist has to extrapolate from them into a dynamic system, to visualise them in transient states which often control the outcome of a reaction and which are not closely represented by the models or drawings. A physical model is the convenient way in which a chemist can become familiar with the three dimensional shape of the molecule he is working with. There are but few experimental techniques which give a direct indication of the spatial relationships of groups, these include X-ray studies and also some sophisticated nmr techniques, and because of this lack a molecular model is routinely constructed.

It would be a mistake to consider molecular models and stereochemistry which they can describe to be the concern solely of organic chemistry; because of the wide range of rings and chains which carbon can form, the study of the shape of these molecules and the influence of three dimensional structure upon the reactions of such molecules, the study of stereochemistry takes a special place in organic chemistry. However, the coordination number of carbon, under ordinary conditions, cannot exceed four whereas inorganic compounds show a range of coordination numbers from 2 to 9. The variety of possible arrangements of ligands around a central atom within this range of coordination numbers recommends the use of molecular models to inorganic chemists also.

With practice, it is not difficult to assimilate the steric relationships associated with the tetrahedral geometry common in organic chemistry, but without the aid of molecular models, detailed examination of the possible interconversions in trigonal bipyramids, now very important in the stereochemistry of phosphorus compounds, or of the many possible seven coordinate geometries, is very difficult.

Mislow (ref. 33*a*) has presented a clear discussion of the limitations of molecular models; his approach can be extended to cover the two dimensional representations discussed in the previous chapter and shows an understanding of the principles of modelling. In summary and translated into our terminology, his argument is as follows.

Measurements of the spatial relationships between groups in a molecule in various configurations can be made with molecular models which abstract and standardise two elements of its prototype, the bond lengths and bond angles of the molecule. The model presents these elements in a clear visual form which can be manipulated following the rules of the conceptual model. Molecular models can not only be viewed from any angle, but can also be strained and generally pushed around; bonds can actually be broken and made. They are, in Bruner's terms apt for both enactive and iconic modelling and moreover have a strong link with symbolic chemical models. Small wonder that they are so popular. The intelligent use of molecular models of course, requires an appreciation of the simplifications involved which are observed particularly in the following circumstances:

(1) Compounds of a given element with the same coordination number will have different bond angles dependent upon the nature of the substituents. Models do not show this. For example, a molecular model would show the bond angles of all compounds containing sp^2 hybridised carbon as the same. The experimentally determined angles

are shown on the formulae in figure 5.15 (ref. 33*b*). Compared with other elements, carbon is regular in the bond angles which it forms and for a given geometry, variations are greater in the case of hetero-atoms, both metals and non-metals.

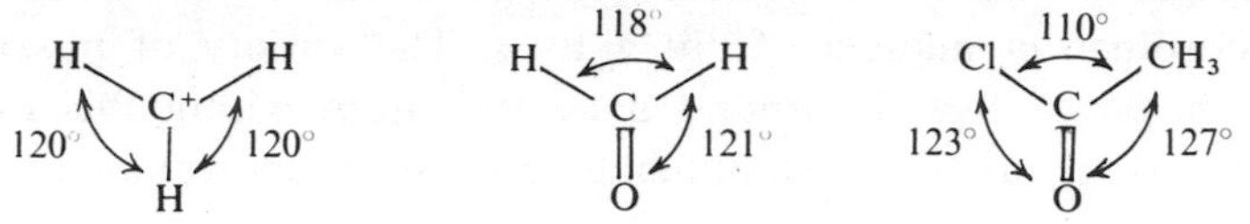

Figure 5.15. Bond angles in sp^2 hybridised carbon.

(2) Bond distances are usually reliable. This fact has considerable significance in relation to the calculation of the relative stability of different possible conformations of a molecule (see below).

(3) An adequate picture of molecular motion cannot be obtained. In this limitation the molecular model is similar to the solution of an X-ray crystallographic study which describes the molecule in one conformation which is stable in the crystal. This problem of depicting dynamic processes assumes further importance in detailed studies of the mechanisms of enzyme catalysed reactions which are discussed in the next chapter. It is, of course, well understood that molecules very often do not react in their most stable conformation.

(4) Many different types of model are available (ref. 34) and the user should be aware of the characteristics of each. It is possible that the use of a different type of molecular model may lead to a different conclusion.

(5) Models are insensitive to the changes in the properties of molecular aggregates with temperature. Temperature not only affects the number of molecules existing (on average) in a particular conformation but also alters the relative energies required for the attainment of different competing transition states and hence the possible courses of reaction.

We can put molecular models to many uses, some of which we have hinted at above. Primarily, they perform the simple iconic function of 'looking-like' their prototype to the best available approximation. In this role, we can examine the model to perceive both the overall shape of the molecule prototype, whether it is long and fibrous or globular and compact, or we can probe more deeply into the stereochemical relationships between groups. The latter can be done statistically, merely to define or measure a relationship, or more deeply in a dynamic form in an attempt to predict the stereochemical course of a reaction, which is the chief function of molecular models in synthesis. Some examples illustrative of these combined applications now follow.

5.4.1 *Reduction by an organometallic reagent*

This example illustrates the choice between possible transition states in the reduction of a carbonyl group which is related to the empirical rules of the conceptual model such as Cram's rule (see chapter 4). In a synthesis of an indole alkaloid, it was desired to reduce and alkylate the intermediate (5.16.1, figure 5.16) (ref. 35). In order

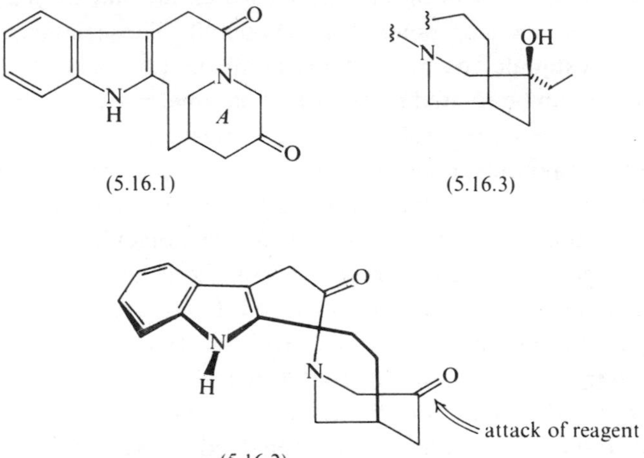

(5.16.1)   (5.16.3)

(5.16.2)

Figure 5.16

to decide which conformation of this molecule was likely to be the most stable and hence to predict the most stable transition state for the reaction, a molecular model was built. This model is illustrated by the structural formula (5.16.2). Considerations of steric interactions between the substituents on ring A from the model suggested that attack of the reagent, ethyl magnesium bromide, was easier from the direction indicated by the arrow on the conformation illustrated to give the product (5.16.3). This structure corresponded to the configuration of the natural product, the goal of the synthesis, and it was predicted that the product predominating in the reaction mixture would be the required compound (5.16.3). However experimental work showed that use of the model over-estimated the stability of the conformation illustrated by (5.16.2) as the reaction products were epimeric. Thus, under the conditions used for the reaction, there was no energy barrier to attack at either face of the carbonyl group.

The models gave an erroneous prediction because the large 9-membered ring in (5.16.2) is probably more conformationally flexible than the model would suggest so that steric interactions around the carbonyl group are relatively non-specific.

5.4.2 *Conformations of metal chelate compounds*

For organic systems it is common to use molecular models to illustrate not only bond lengths and bond angles but also the existence of steric interactions between groups which are not directly bonded together (so-called non-bonded interactions). These interactions can be used to estimate the most stable conformation, as in the previous example, or can be related to empirical parameters which measure the energies of interaction (ref. 36). There exist also a large number of inorganic compounds, chelate complexes of metal ions, which contain rings of various sizes, usually between three and six members. In these compounds too there is a wide scope for the molecule to adopt distinct conformations, and an understanding of the relative stability of the possible conformers may be obtained from examining molecular models of the compounds. Thus, for six-membered ring chelates, models suggest the possibility that three stable conformations exist (figure 5.17), the chair, the boat, and the skew-boat conformations. Models of such compounds have been used to estimate the distances between the substituents on the ring and on the metal and these values, which as Mislow pointed out, are usually reliable, were fed into equations deduced from theory which allow calculation of the relative stability of the conformers. Not surprisingly, the same qualitative conclusion is usually reached as one would predict by inspection of the model. For example, the bis(ethylenediamine)sarcosinatocobalt (III) ion has two chiral centres, the metal itself, and the nitrogen atom, and it can therefore exist, theoretically, in four possible stereoisomeric forms. However, only two of these possibilities have been isolated (ref. 37 figure 5.17), compound (5.17.1) and its diastereoisomer with the opposite configuration at cobalt. Calculations based on measurements of bond lengths from a molecular model surprisingly showed that (5.17.1) was about 42 kJ mole^{-1} more stable than its nitrogen epimer (5.17.2) because of the extremely large non-bonded interaction between the methyl group of (5.17.2) and one of the ethylenediamine–cobalt rings. This implies that in similar cases, it should be possible to prepare complexes of predetermined stereochemistry by taking advantage of the inherent instability of certain stereoisomers, a prediction which has been realised in practice (ref. 38).

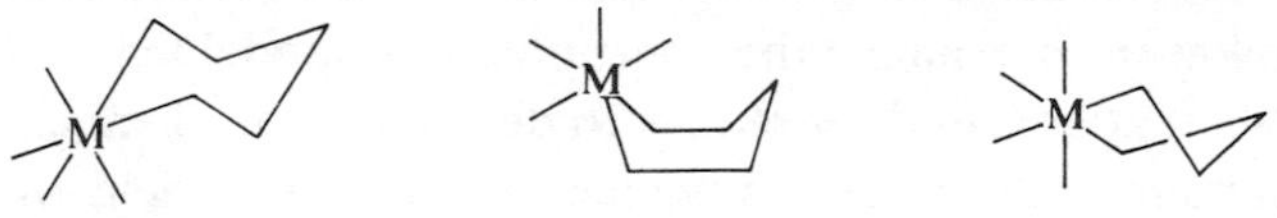

Chair   Boat   Skew-boat

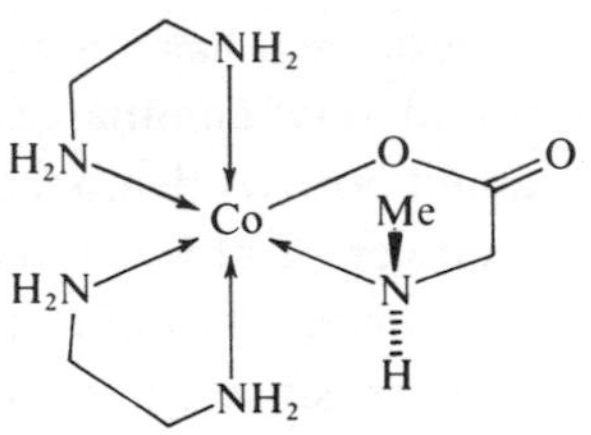

(5.17.1) 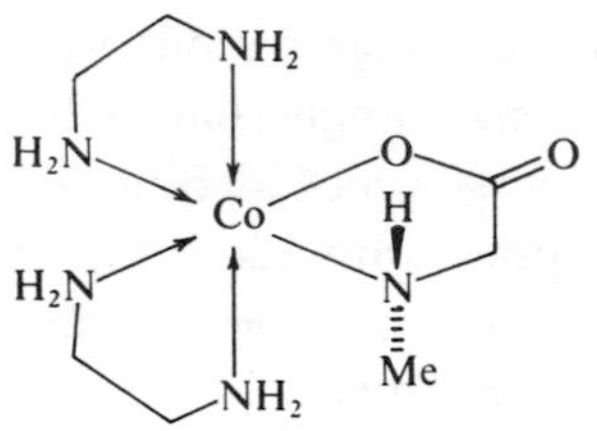   (5.17.2)

Figure 5.17. Conformations of metal chelates.

These examples show that even with careful understanding of molecular models it can be difficult to estimate by inspection which of several competing reaction paths will predominate. That is, in kinetically controlled reactions, which transition state is most stable, and in thermodynamically controlled reactions, which is the most stable product. Thus the greatest problem of the conceptual model and its associated techniques is again encountered, namely how can one estimate with chemical accuracy the energies of closely similar species under defined conditions without going to the expensive lengths of a rigorous theoretical calculation. In many cases, the qualitative treatment offered by the conceptual model and inspection of the molecular models is sufficient and correct predictions are made. However, as these examples illustrate, predictions made in confidence can be proved wrong by experiment or equally can be reinforced in a striking manner. One should never refuse to attempt a reaction, therefore, just because the inspection of molecular models makes it look unfavourable, nor should one be surprised if the unexpected occurs. The complete exclusion of a dynamic element in molecular models is a sufficient abstraction to make the prediction of reaction pathways a difficult extrapolation.

5.5 Structure determination and physical methods

In the previous section, we dealt with one of the most important supporting techniques of the chemist, the use of molecular models. We now turn to examine the contribution of modelling to the physical methods that are used nowadays to determine the chemical structures of compounds. The great advances made in this field over the last twenty years have in a large part been due to the use of physical methods which enable structures to be determined with great rapidity using extremely small quantities of material. The necessity for lengthy chemical degradations to prove structures proposed has been reduced correspondingly but it has not been completely eliminated. Often confirmation of a proposed structure is sought both by degradation and by total synthesis. As an example of the power of physical methods, the determination of the structure of the insect juvenile moulting hormone (figure 5.18) was a striking success since only 365 μg of material painstakingly isolated from many thousands of insects (ref. 39) was used.

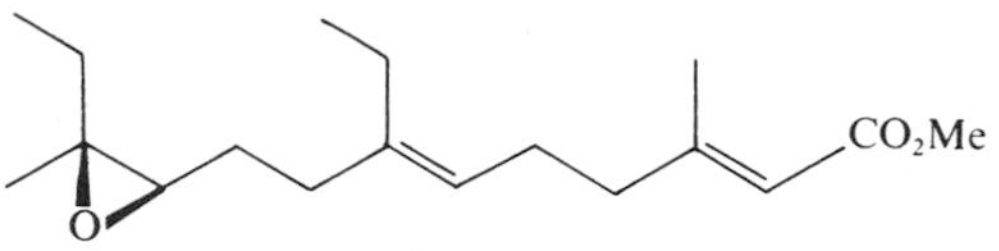

Figure 5.18. The structure of juvenile hormone.

Why have these physical methods proved to be so powerful? The answer can be viewed in terms of a series of well defined modelling transforms applicable under closely controlled conditions (in other words a fully defined system is employed). This applies to all of the commonly employed spectroscopic methods including infrared, electronic, and nuclear magnetic resonance spectroscopy, mass spectrometry, and also to the inorganic technique of magnetic susceptibility measurements. Most of these methods have a firm theoretical basis, indeed the physical measurements themselves are often the parameters used to test the predictive powers of theoretical treatments. In some cases, particularly nmr and esr spectroscopy, it can be of great value to reverse the process and to use the proven theoretical treatments to predict spectra of interest or to analyse spectra which are too complex for interpretation by simpler more empirical methods. However it is not in the theoretical background that the great practical value of physical methods is realised, the value is that important structural features in a molecule can usually be recognised by a characteristic value of the physical property measured. For example, in the infrared spectrum, an ester carbonyl group is usually indicated by an absorption at 1725 cm^{-1}, a tertiary methyl group absorbs in the proton magnetic resonance spectrum at a chemical shift of about 1 δ (9 τ) and a metal ion with one unpaired electron shows a magnetic susceptibility of 1.73 Bohr magnetons. The theoretical reasons for these results are well understood but they need not be considered in detail when the techniques are applied in the laboratory. The fact that spectral properties only vary within a narrow range for a particular structural element or functionality allows collections of standard data to be assembled. Measurements are made under standard conditions and the properties are usually invariant within the time scale of the measurement and therefore one can correlate the spectra of known compounds with those of unknown compounds drawing the general conclusion that like groups will show the same spectroscopic properties. However one must always pay heed to the possibility that a combination of a number of unexpected factors may conspire to give an unusual result. The method of correlation is usually applied following tables of data or memorised typical values such as those quoted above. In more complex cases, a direct analogy with another compound in the literature may be used; this is clearly an extension of the modelling by analogy or 'off-the-peg' models discussed earlier. This approach is successful almost without exception and the reasons for this in terms of our analysis of modelling are not hard to find. The measurement of spectra is a simplified case of the reacting system considered earlier. In a spectrometer, we normally do not observe reactions, hence reactant–reactant interactions, usually the most difficult to evaluate when constructing a model, are eliminated. Solvent–solute interactions are of course observed but these are either eliminated for the purposes of the test by the use of standard conditions or they can be used to supply further information about the system under test. Temperature changes can be treated similarly. Thus the only interaction observed is the one of interest, the interaction between the probe of the method and the substrate under test. In this way we find one of the closest correspondences between prototype and model, and the form of the model, a collection of tables, is one of convenient simplicity. Often the prototype may be a single functional group or atom rather than the whole molecule and the physical methods allow us to probe the properties of these groups without the need to introduce the uncertainty of other reagents.

So far, we have considered in general modelling terms the reasons for the great success of physical methods in structure determination. Examples are numerous and the modelling can range from the correlation approach that we have just discussed to theoretical modelling involving quantum mechanical calculations as we shall now see.

5.5.1 *Electronic spectra*

Correlation of the physical properties of an unknown with standard model compounds is the quickest and easiest method of applying spectra to structure determination. In many cases, the correlation is made directly with the compilations of literature examples, but, on occasions, specific model compounds are made for comparison. A case in point is the antibiotic terramycin, a tetracycline that we have met before (chapters 3 and 4). The ultraviolet spectrum of terramycin (5.19.1) was not recognised as typical of any known class of compound but from other data, it was suspected that a chromophore of the type occurring in the model compound (5.19.2) was present. The spectral data for the model compound and unknown prototype are quoted in figure 5.19. It can be seen that although the total structures of the two compounds are vastly different, the spectra are similar enough to establish a positive correlation (ref. 40). Provided that there is no interaction between isolated chromophores or functionalities, the method of correlation is very reliable. The lack of an expected correspondence often indicates the operation of unpredicted effects, another common feature of models, which can on occasions be turned to advantage (see below).

Because electronic spectra show the differences between electronic energy levels of a molecule, they are well understood theoretically. This is true both for organic and inorganic molecules. All the treatments are sensitive to the fact that molecular geometry plays a great part in dictating the spectra. Cobalt(II) complexes are a good example of this since they exist in two major geometries, tetrahedral (blue) and octahedral (pink). Typical cases are shown in figure 5.19. However these are not the only possible geometries and in special cases where the ligands control the geometry through their structure, square planar and trigonal bipyramidal structures are possible. Such spectra are diagnostic of a particular coordination geometry for cobalt and they are the primary tool in determining the structure of a complex (ref. 41).

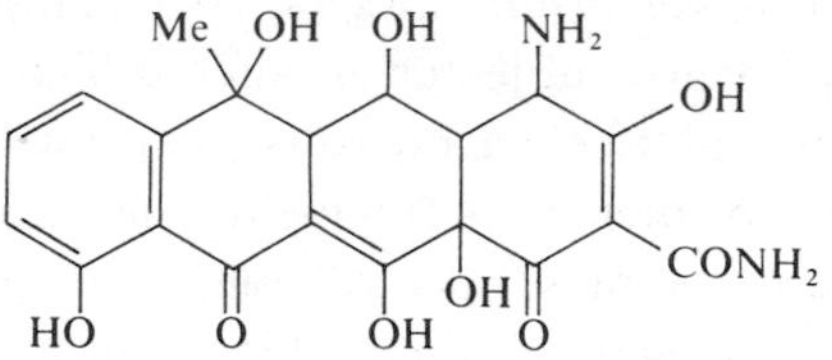

λ_{max} 267 nm ϵ 21 600
360 nm ϵ 12 500

(5.19.1)

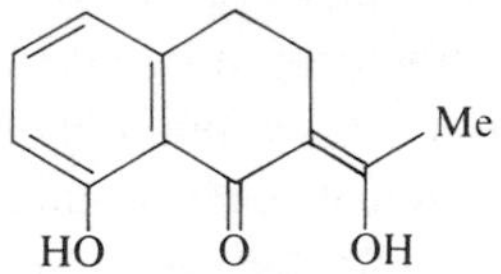

λ_{max} 260 nm ϵ 5700
345 nm ϵ 12 500

(5.19.2)

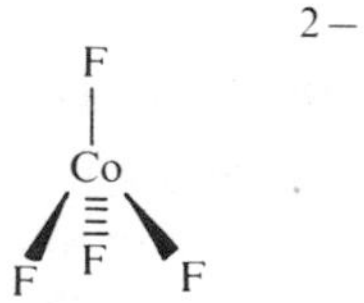

Tetrahedral (blue)

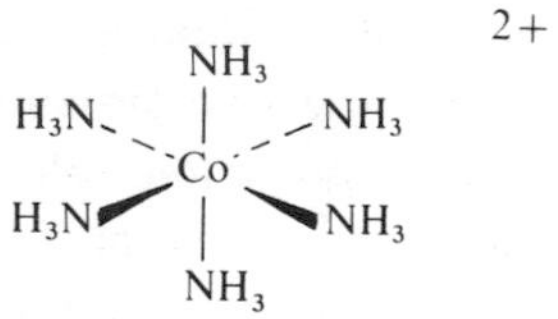

Octahedral (pink)

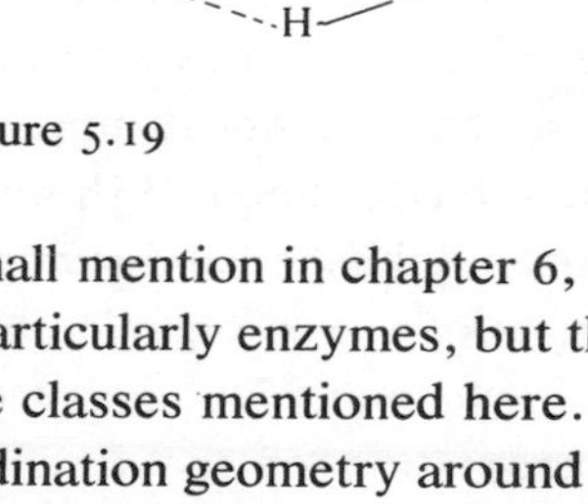

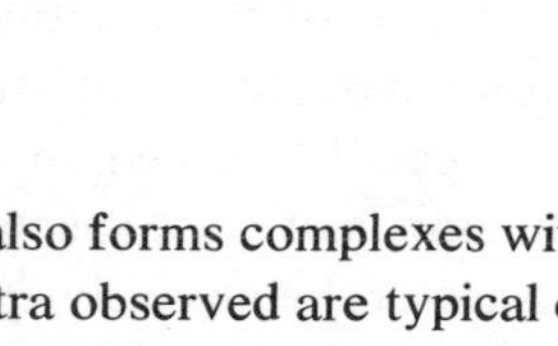

Figure 5.19

As we shall mention in chapter 6, cobalt also forms complexes with proteins, particularly enzymes, but the spectra observed are typical of none of the classes mentioned here. The lack of correlation implies a novel coordination geometry around the cobalt and this geometry is of considerable significance in understanding the mechanisms of action of the enzymes concerned.

5.5.2 *Mass spectra*

Most of the forms of spectroscopy that we discuss in this chapter depend on the interaction of electromagnetic radiation with the substance under study. No chemical reaction occurs. On the other hand, when we measure a mass spectrum a drastic chemical process takes place. Here the molecules are subjected to such a high energy (commonly a beam of electrons at 70 electron volts) that not only do they ionise but they are also broken into fragment ions. Clearly a chemical reaction takes place in the mass spectrometer and a possible way of interpreting such a reaction is to use the analogy of ground state reactions in solution. In this way the 'fish-hooks' that are commonly used to analyse mass spectra are equivalent to the 'curly arrows', the conceptual formalism used to provide a dynamic element in descriptions of solution reactions.

But the two situations are very different. The energies and interactions present in solution reactions are very different from those in the mass spectrometer. In the mass spectrometer solvation is absent and the ions formed contain one electron less than the so-called analogous species of the former. Indeed very little is known about the structure of ions in the mass spectrometer and so we have scant information to base any model upon. This problem of similarity between mass spectrometric and solution reactions has been discussed in detail by Bentley and Johnstone (ref. 42a). However, as they point out, there is nothing wrong in using a pictorial representation of fragmentation in the mass spectrometer as an aid to structure determination as long as it is not held that the drawing is a realistic representation of what is going on. The over-reliance on 'curly arrows' was criticised in the discussion of the reacting system in chapter 4 and since so little is known about the structure of ions in the mass spectrometer, very little faith can be placed in arguments based upon 'fish-hooks' alone. Thousands of mass spectra have been published and summaries of the most useful empirical generalisations of use to organic chemists have been compiled by Djerassi, Budzikiewicz and Williams (ref. 43).

For example, a pictorial representation of the fragmentation of 1-benzyl isoquinolines (e.g. norlaudanosine, chapter 6), a common structural type of alkaloid, is shown in figure 5.20.

In conventional chemistry, the drawing of such structures implies a great deal. Thus we would expect a derivative of quaternary nitrogen, $\overset{\oplus}{\underset{/\,\backslash}{N}}$, to undergo an elimination reaction just as the ion represented by

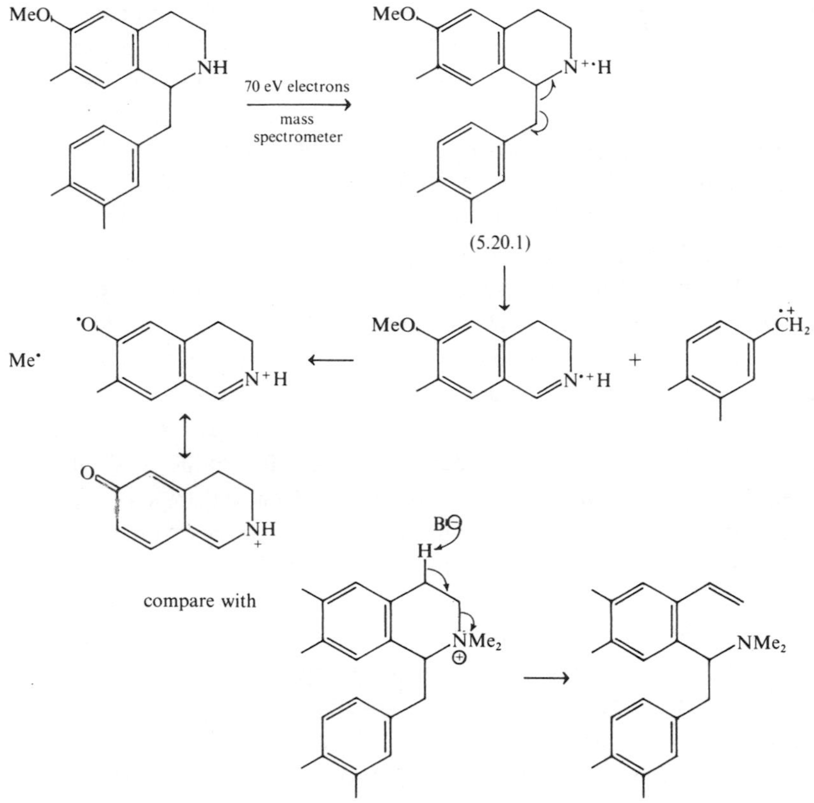

Figure 5.20. A representation of the fragmentation of 1-benzyl isoquinoline.

(5.20.1) does here, but we dare not press the mechanistic analogy further for the reasons stated above. In a case such as this which has two electron-rich aromatic rings, it is highly probable that the initial loss of an electron to form the molecular ion is from one of these two rings and not as depicted from the nitrogen atom (ref. 42*b*). As a conceptual rationale typical of the most abstract form of the conceptual model of chemistry, 'fish-hooks' and localised charges provide a tidy means of organising and correlating the mass spectral behaviour of many classes of compounds, but it is important that one should not assign to 'fish-hooks' more than the status of a pictorial empirical correlation model. Fortunately, Djerassi has carefully defined the large number of possible fragmentations in terms of the 'fish-hook' model and his books (ref. 43) based upon this system provide ready access to a large body of precedent. Because of the similarity with curly arrow symbolism in ground state chemistry, a strong temptation exists to allow the fish-hook model the same freedom of extension. However attractive, one must not allow a model to be separated from its frame of reference, otherwise not one valid prototype–model transform exists.

5.5.3 *Nuclear magnetic resonance spectroscopy*

We have seen how models of the correlation type, the empirical type and the conceptual type can be used as aids in the application of physical methods to chemistry but it is also possible to apply rigorous quantum mechanical procedures to assign the origin of peaks in a spectrum to their source groups. It is a relatively simple matter to calculate the electronic spectra of both inorganic and organic molecules and the small number of bands obtained are usually easy to assign. However, a nuclear magnetic resonance spectrum (nmr) can contain many lines, often grouped as multiplets, and the correct assignment of each line cannot be made by inspection. Since the quantum mechanical theory of nmr was developed immediately following the discovery of the phenomenon of magnetic resonance, it has become possible to describe nmr spectra in terms of a quantum mechanical model which can be used with the aid of computational techniques to analyse complex nmr spectra and to obtain from them the information which the chemist requires. This analysis is a specialist exercise in complex cases but books aimed at helping the chemist to appreciate what can be done have been written as the following quotation from Abraham shows (ref. 44): 'the analysis of an nmr spectrum may be considered not as an exercise in quantum mechanics, nor even an end in itself but primarily as a necessary service before the chemistry can start...' The most common applications of such analyses are for the nmr spectra of protons. In order to set up the equations for analysis it is simply necessary to translate the nmr parameter associated with each nucleus, the chemical shift, with its associated interaction with another nucleus, the coupling constant, into equations and then follow standard quantum mechanical procedures (ref. 44). When these equations can be written down explicitly, calculations of chemical shifts and coupling constants can be carried out at the desk using a pocket or desk calculator, but in more complex cases, iterative computer programmes must be used (ref. 44). Chemically, it is more useful to be able to calculate from the positions of the lines in the spectrum the chemically meaningful parameters, the chemical shift and coupling constant. However the reverse process has been carried out and a vast library of calculated nmr spectra is available which can be used to correlate unknown spectra. The coupling constant, *J*, is particularly useful in determining stereochemical details of organic molecules because it has characteristic values corresponding to defined relations between the interacting nuclei. A theory of the coupling of nuclear spins was developed from valence bond arguments (chapter 3) by Karplus and his theory predicted the magnitude of coupling constant to be expected for a given geometry of a molecule (ref. 45). In particular he predicted that the coupling constant should vary with the dihedral angle between the C–H bonds which include the two interacting protons. A common structure which contains protons in a fixed relationship is the *trans*–decalin system found as part of the steroid skeleton. The predicted values of the coupling constant between two vicinal axial protons is 8–13 Hz in such systems and for an equatorial proton interacting with a vicinal axial proton 3–5 Hz. These values have been confirmed in practice and now can be considered as firm criteria for deducing stereochemistry of rigid systems.

As an example of the application, let us consider the cyclisation of squalene epoxide (5.21.1) under acidic conditions which, as we shall discover in the next chapter, is a good model of the biosynthesis of steroids and triterpenes. Under the conditions used for this experiment (ref. 46), the major product was the bicyclic compound (5.21.2) and the problem was the stereochemistry of the hydroxyl group in the product. The solution to the problem came conclusively from the nmr spectrum of the epimeric alcohol which was prepared from (5.21.2) by oxidation and reduction with sodium borohydride. The spectrum of the epimeric alcohol (5.21.3) showed a quartet centred at 3.12 δ corresponding to H_X from which the coupling constants of H_X with the vicinal hydrogen atoms H_A and H_B could be determined. The values $J_{AX} = 10$ Hz and $J_{BX} = 4.5$ Hz are typical of an axial–axial interaction and an equatorial–axial interaction and hence, alcohol (5.21.3) can be assigned the configuration shown and the product (5.21.2) the epimeric configuration, which is the opposite of that found in the natural products.

Many nuclei other than protons have observable nmr spectra and the study of such spectra has allowed not only structure determination to be carried out (e.g. ^{31}P compounds) but also the mechanisms of

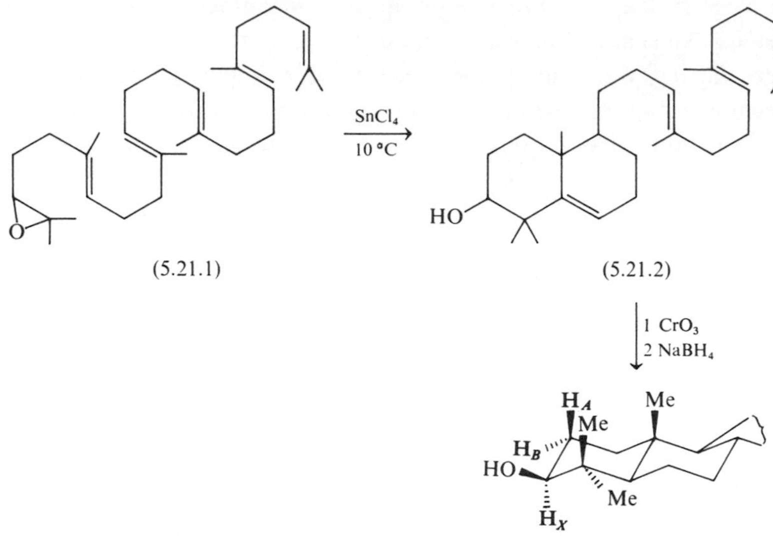

Figure 5.21

reactions to be studied. This can, of course, also be achieved with protons, and the study of ligand exchange processes using ^{57}Co nmr or of redox processes using ^{63}Cu shows the wide range of applicability of the nmr technique. Reference 47 discusses these and other examples of a correlation type model applied in nmr spectroscopy in detail.

5.5.4 *Infrared spectroscopy*

As our last port of call in our travels through modelling applications in the more practical aspects of chemistry, we shall turn again to our embarkation point, organic synthesis. The examples which we have just considered emphasise the reliability of physical methods in chemistry, but once in a while, things do not fit the well-expected pattern. Stork came across such a problem during his synthesis of camptothecin (ref. 48) which is interesting from the point of view of synthetic strategy as well as spectroscopy. By an unambiguous, carefully developed synthetic route, the intermediate lactone (5.22.1) was prepared and all the spectral data but the infrared spectrum were in good agreement with the proposed structure (5.22.1). However, the γ-lactone showed an absorption at 1802 cm^{-1} which is about 20 cm^{-1} higher than one would expect for this functional group (ref. 49). Since infrared absorption is associated with the stretching and bending vibrations of bonds, Stork reasoned that this lactone carbonyl group

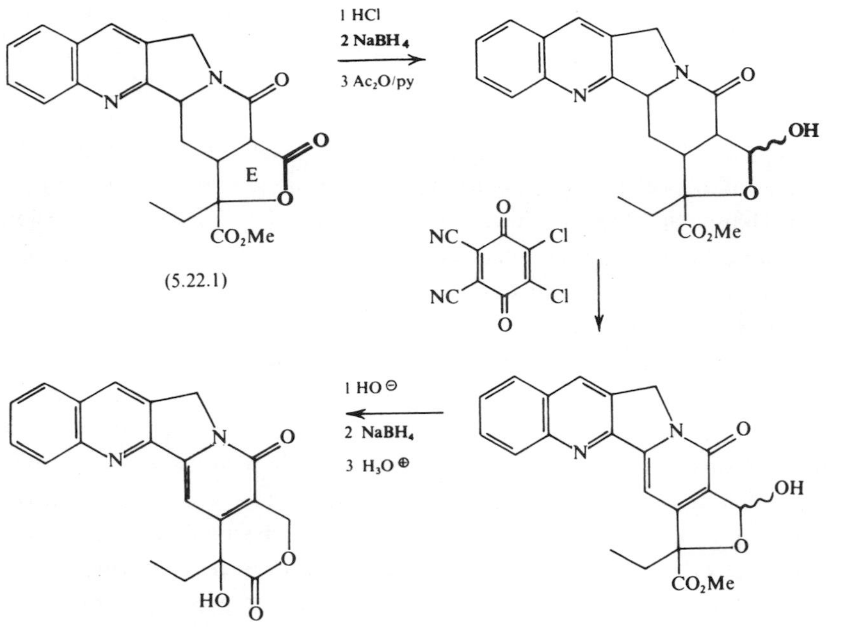

Figure 5.22. Modifications to ring E.

had more than the usual double bond strength for a lactone and was instead more like a ketone. If this were so, then the lactone should be reducible by sodium borohydride, a reagent to which normal lactones, but not ketones are usually inert. This conceptual analogical argument typical of modern organic chemistry proved correct and the completion of the synthesis required only a few simple laboratory reactions.

5.6 Conclusion

It has been the aim of the last three chapters to present a view of a wide range of paper and practical chemistry within a framework of recognisable transformations between models and prototypes. We have always been able to make a clear definition of at least two of the three vital components of a model even though the rigour and sophistication has varied greatly. Thus we have discussed cases in which the prototype and transformations are rigorously defined, as with the physical methods, and also the other extreme in which the model consisted of a set of axioms related to the prototype, whose properties were known, by a process of parametrisation and calculation, as in quantum mechanics. Both of these types have shown characteristics of reliable models. We have also experienced considerable difficulties when the nature of the transformations was obscure despite the fact that many characteristics of the prototype and model were clear. This is typical of the everyday application of what we have called the conceptual model of chemistry to reacting systems of many kinds. What happens, then, when even the prototype is poorly defined? Dare one hope to construct viable models? In fact in two very distinct cases, the answer is a very firm yes, as the remaining chapters of this book will show. The first case that will be considered is the rapidly growing area of chemistry which borders upon biochemistry. Here, chemists of every background are active, not only organic chemists as one might expect, and they are all attracted by the possibility that a model system may predict and explain the behaviour of their prototypes, which are natural macromolecules and their reactions. The second case is the contrasting activity of the use of models in the design of industrial chemical processes, which now include even the production and use of enzymes. In neither case can we define our prototype precisely before we begin modelling, but after modelling, we should be able to define it better.

6

MODELLING OF BIOLOGICAL SYSTEMS

6.1 Introduction

The scope of models of biological systems. Perhaps the most enduring and challenging problem for science is to understand the many different processes that are integrated into living organisms. It is a problem which has been tackled by many scientific disciplines from medical to physical, and for chemists it takes on a particular significance since their discipline provides a frame of reference within which medical, biochemical and physical approaches can find common ground and an opportunity for discourse. For the chemist two distinct problems arise at the molecular level – firstly the determination of the structures of the molecules that comprise living cells and their place in biological synthesis, and secondly the interpretation in mechanistic terms of the pathways through which cells construct and then transform again the molecules that they need. Structure determination involves primarily the study of the pathways of reaction which the biosynthesis of natural products follows, and this leads directly into the second task which is chiefly concerned with the proteins which catalyse the biosynthetic and, of course, the biodegradative reactions. An extension of these problems is the study of the interlocking control mechanisms which regulate the metabolism of entire organisms. In this chapter, we will examine the contributions of model systems to our understanding of these problems in terms of the patterns of modelling discussed in the previous chapters. Model compounds have found their widest use in the study of reactions catalysed by single enzymes and accordingly, this field will feature most prominently in the discussion.

Although today we look upon the interdisciplinary nature of chemical studies of biological systems, particularly enzymes, as novel, in the late nineteenth and early twentieth centuries, when the recognition of enzymes as catalysts was emerging, the scientists actively working in this field were men who also had a profound influence upon 'classical' chemistry, for example Pasteur, Emil Fischer and Willstätter. A glance at the contents of *Hoppe–Seylers Zeitschrift für Physiologische Chemie* will show that a formal distinction between chemistry and biochemistry was not contemplated. Although the formal classification of simple organic compounds according to functional group (see chapter 3) was clear, there was no unifying theoretical basis which could predict the properties of functional group combinations in complex molecules, nor were the structural relationships between the different members of well-known classes of natural products such as terpenes and alkaloids discerned. Had the analogical nature of chemistry, particularly organic chemistry, been apparent then, the time would have been ripe intellectually for the development of biological model systems. However, chemical systematisation arrived half a century too late to save the study of enzymes and proteins from battling not only against the intrinsic elusive nature of the complex natural molecules, but also against the prejudice of 'traditionalists'. The successes of research in the field of biosynthesis have helped greatly to make the chemist aware of the potential in the field of biology of his way of looking at things. This awareness has also more recently spread to the inorganic chemist with the discovery of the functional importance of metal ions in enzymic catalysis. The proliferation of titles for publications such as bio-organic chemistry and molecular biophysics is evidence for the point.

6.1.1 *The conceptual basis for models of biological systems*

It is remarkable that despite the progress made in the last twenty years, many chemists instinctively react to an enzyme as if it possessed some magical qualities. Such a reaction may have had some justification in the era before protein structures became determinable by X-ray crystallography but today it is manifestly absurd. Chemistry is the fundamental science which deals with the ways in which molecules interact with each other and encompasses all species from the size of a hydrogen atom to a large protein. But the rules of the game remain unaltered whatever the nature of the players may be. A carbonyl group in any molecule will always be susceptible to nucleophilic addition reactions provided that the appropriate experimental conditions can be found. If we do not observe what we expect, it need not be because the reaction cannot occur, it may well be that we simply had the wrong reaction conditions.

It so happens that many biologically important molecules including enzymes have rather limited chemical stability and therefore their reactions cannot usually be observed under the traditional conditions of the chemical laboratory. The lack of appreciation of the chemical nature of proteins, enzymes and other biological macromolecules has thus an experimental basis but it also stems from neglect of an underlying philosophical principle that runs through this chapter.

When we consider what might happen in a chemical reaction, we look at each reactant in turn and try to generate a mental picture of the chemical character of each. Then we look and see how the two might react together following the lines of the chemical character we have perceived. The more complex the molecule, the more difficult our task becomes and the complexity increases when we consider biological systems.

The chemist's analogical approach to biological systems. Fortunately, however, most biochemical reactions involve relatively small molecules reacting under catalysis by the enzymes. We can understand a great deal of what is going on by focussing our attention upon the inherent chemical reactivity present within the small reactant molecule. For example, the biosynthesis of thymidilic acid (figure 6.1) from urydilic acid can be interpreted formally as a complex analogue of the well-known reaction, electrophilic aromatic substitution. Urydilic acid is composed of uracil and ribose-5-phosphate and in this reaction, the chemically relevant species are a methylating agent and uracil. The ribose phosphate can be regarded simply as an 'R' substituent. If we consider the intermediates that arise when uracil undergoes an electrophilic aromatic substitution either by MO or VB methods, we shall see that electrophilic substitution is most favoured at the 5-position. Accordingly, uracil undergoes the typical aromatic substitution reaction, nitration, readily with concentrated nitric acid and it is substituted at position 5. Similarly we see that our biosynthetic transformation involves methylation at position 5 and we would therefore expect that some electrophilic, one carbon fragment is also a participant in the reaction. Biochemists have shown that the coenzyme N-5,10-methylenetetrahydrofolic acid (6.1.1) is required. When protonated, this complex molecule contains essentially an $R–CH_2^+$ species disguised as a relative of the well-known electrophile, a Mannich base. Notice that the bulk of these large molecules are not concerned with the reaction itself. Similarly only a small part of the catalyst enzyme molecule may be directly interacting with the substrate.

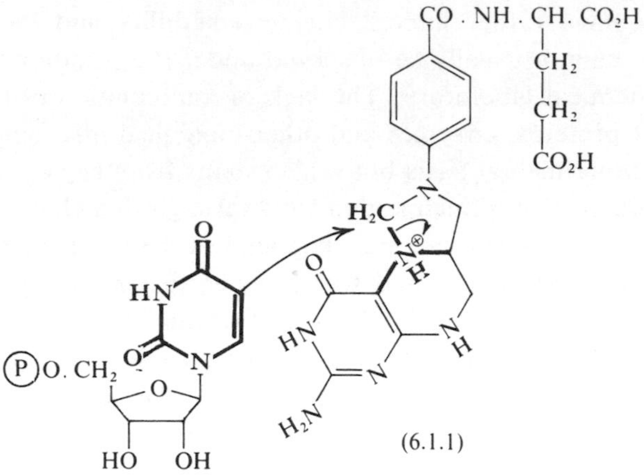

Figure 6.1

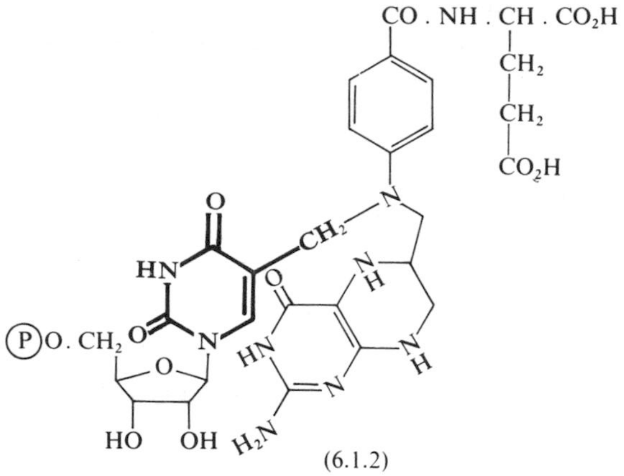

Now we have carried out the substitution upon urydilic acid but we have obtained a derivative of thymidilate that is substituted by the remains of our methylating agent. A further transformation is required in order to obtain the final product, thymidilic acid. As far as the uracil derived fragment is concerned, a substituent must be replaced by a hydrogen atom, that is to say a reduction must occur. This reduction is something new. There exist no close analogies for this reaction and so the innate chemistry of intermediates such as (6.1.2) must be investigated. The study of simpler model compounds is a suitable

Urydilic acid   Thymidilic acid

(6.1.3)

A Mannich base

approach and recently results of such work have begun to appear (refs. 1, 2).

It may well be that the plausible formal analogy that we have described may prove to be incorrect in the light of further experimental results. Most probably we have oversimplified especially because we have knowingly not considered how the catalyst enzyme might interact with the substrate during reaction. Chemically, the most advantageous point in the reaction (6.1.1) → (6.1.2) for the enzyme to intervene is at the stage of the positively charged intermediate analogous to (6.1.3).

Here, the energy of the reaction is close to its maximum and stabilisation of such an intermediate by an electron donating group would
offer a powerful mechanism for catalysis.

The biological reaction is, of course, catalysed by enzymes and our
understanding of the chemistry of the other reactants can be used to
tell us something about the enzyme. Although the detailed mechanism
of the enzyme catalysed reaction is unlikely to be the same as the
mechanism of the chemical reaction, the enzyme must promote the
reaction by employing the very chemical properties that we have
discussed. The enzyme can do nothing else but transform the substrate
according to the substrate's innate reactivity. This very important fact
is valid whether we are simply interpreting a reaction in our familiar
conceptual chemical terms or whether we are consciously designing,
constructing, and operating a model. All of these problems are the
concern of this chapter. Let us now turn our attention to consider the
sort of experimental evidence which can be obtained in biological
systems, the evidence that provides a basis for modelling.

6.1.2 *The experimental basis*

We know that modelling of poorly understood systems is
possible and can be useful, but chemistry tends to work with models
rich in interaction and these are difficult to construct unless we have
a good idea of what the prototype actually is. And so the first experimental task is to define the prototype as far as possible.

Biosynthetic studies and the tracer approach. The first chemical approaches to biological systems came from the study of biosynthesis
and one commonly used approach employs isotopically labelled molecules. The chemist postulates a possible sequence of precursors for
the natural product in question based upon a critical chemical examination of its structure, a modelling process which has become axio-
matic in this field. One or all of these precursors may then be
synthesised containing a radioactive or other isotopic label at a fixed
position in the molecule and fed to an organism (plant, animal,
microorganism or to an extract from the organism) which biosynthesises
the natural product being studied. The organism is killed and the
metabolites in question are isolated, purified and carefully checked
for label content. The position of the label in the metabolite molecule
is then determined by chemically degrading the metabolite to recog-

nisable fragments. If the label is in fact found in the position indicated by the postulated biosynthetic pathway, then the compound that was fed, in principle at least, has been proved to be a precursor of the natural product in the organism studied. Sometimes a label is found in an unexpected position; in this case, the original biosynthetic scheme must be modified or discarded. For example, in 1917, Robinson (ref. 3) recognised that substituted phenylethylamines were very probable precursors of the benzyl isoquinoline group of alkaloids and by such experiments as illustrated by figure 6.2, his postulate has many times been proved correct (ref. 4). Today, through feeding experiments using well-chosen precursors, it is possible to elucidate many subtle details of biosynthetic processes, including stereochemistry.

The reader will find little difficulty in relating the train of thought in this chapter to the principles of modelling that were discussed in chapter 2. He may observe, however, that we say little about the physical aspect of the problem, the actual experimental technique that

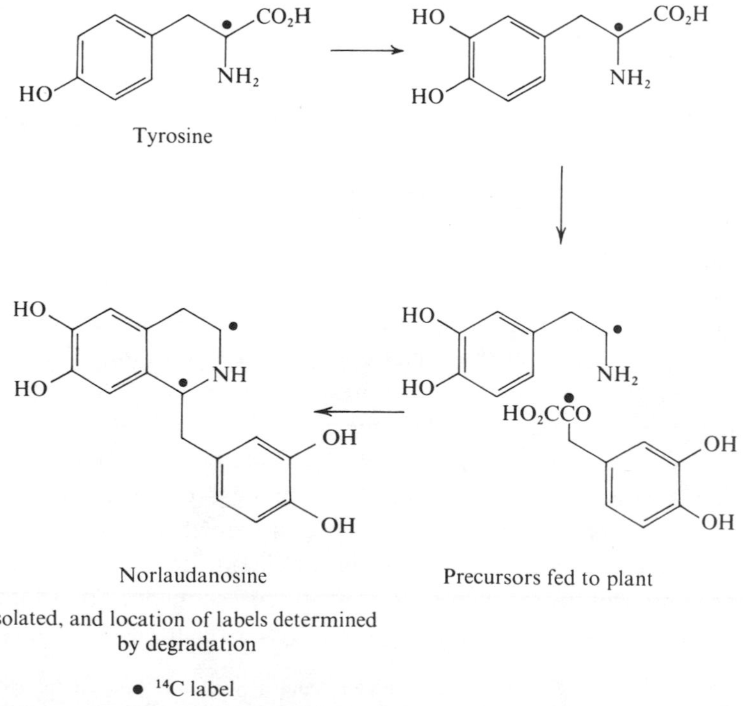

Figure 6.2. The biosynthesis of norlaudanosine, a benzyl isoquinoline alkaloid (ref. 5).

corresponds to making and running the model with its skills and know-how that make all the difference between success and failure and, indeed, between safe and dangerous practice. The chemist will, of course, deal with these matters during the normal course of his work and it is not the object of this book to offer another traditional presentation. Many of the review articles and undergraduate texts cited in this book offer helpful discussions of practical techniques.

The chemical nature of enzymes and its implications for models. Enzyme molecules are very large (M.W. = 10^4 to 10^6) but the forces, such as hydrogen bonding, which hold them in reactive conformations are weak compared with those that operate in most covalent bonds. This results in a high degree of instability under the conditions that are required for most synthetic and degradative reactions in the laboratory. Thus, the complex and unstable nature of enzymes makes a direct assault upon their chemistry difficult in the majority of cases; the point at which these direct studies break down is the cue for modelling.

As we have mentioned, many chemical transformations catalysed by enzymes are readily recognisable as the biological equivalents of well-known reactions such as amide hydrolysis (catalysed by a peptidase e.g. chymotrypsin, E.C. 3.4.4.5)* (figure 6.3), aldol condensation (catalysed by malate synthetase, E.C. 4.1.3.2) (figure 6.4) and by elimination and hydration, e.g. of water (catalysed by fumarase, E.C. 4.2.1.2.) (figure 6.5).

Such is the complexity of enzymes and so incomplete our understanding of their working that a single model that would comprehensively represent the behaviour of an enzyme cannot be formulated and in practice models concentrate upon one selected aspect of the catalytic process, as will be seen.

A model cannot be conceived, let alone be constructed, until some fundamental information about the process to be modelled is available. The system must be defined and here this necessitates that the reaction to be studied must be clearly identified (as might be done by feeding an intermediate in a biosynthetic sequence for example) but it is often helpful with enzyme catalysed reactions to know also the

---

* The E.C. number is taken from a systematic numerical catalogue of enzymes according to the reaction they catalyse. The catalogue is compiled by the Enzyme Commission of IUB.

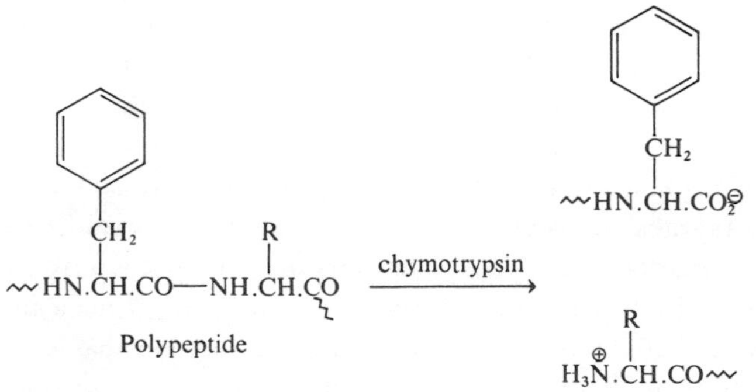

Figure 6.3. Protein hydrolysis by chymotrypsin

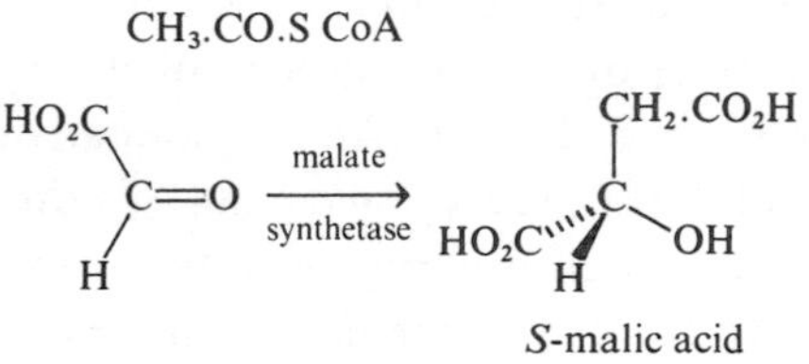

Figure 6.4. An aldol condensation catalysed by malate synthetase.

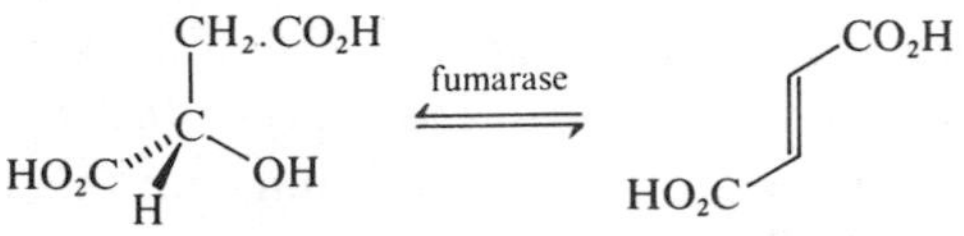

Figure 6.5. The elimination of water from (S)-malate catalysed by fumarase.

results of kinetic or inhibition experiments, the spectra of the enzyme and the possible involvement of cofactors. For example one might wish to establish whether the reaction that is catalysed by the enzyme shows a dependence of rate upon pH. Or one might ask: do reagents which attack sulphydryl groups such as iodoacetate seriously inhibit the activity of the enzyme? Is the presence of a particular metal ion an essential prerequisite for reactivity? The answers to these and many other similar questions provide information about the *active site* (ref. 6) of the enzyme, that is the region of the surface of the enzyme at which transformation of the substrate occurs.

The existence of active sites on or near the surface of an enzyme has been very thoroughly substantiated by experiment. The composition and form of the active site is dictated by the conformation which the polypeptide chain is forced to adopt by its own internal constraints, largely hydrogen bonding between peptide bonds, and covalent disulphide bridges between cysteine residues in the chain. Sometimes external cofactors such as metal ions also impose constraints. The composite effect of all these forces is to bring together a number of amino-acid side-chains to provide a location that is capable of binding the substrate to it; a binding site often can only bind one specific substrate (for example malate synthetase will accept only acetyl coenzyme-A and glyoxalate as substrates). Furthermore, the interactions of an intimate association of side-chains may modify greatly the individual properties of substrate or reactive groups of the active site from what they would be in isolated solution. An enzyme can thus exert a microenvironmental control over the reaction it catalyses through the various reacting groups, a fact which has great significance for model studies; thus the internal environment of the model and the prototype as well as their interactions with their respective external environments are in need of particular attention from the modeller.

Chemical modifications of enzymes – Papain. A simple example will help to illustrate what can be deduced from preliminary chemical experiments and further cases are discussed throughout this chapter. Papain (ref. 7) catalyses the hydrolysis of amides and is a relatively small protein having a molecular weight of 23,000. The following experimental results have been obtained by studying the reactions catalysed by the enzyme and by reacting the enzyme with other chemical reagents.

(1) Mercurial reagents prevent the enzyme from hydrolysing amides.

(2) Treatment of the enzyme with iodoacetamide yields an inactive enzyme which upon total hydrolysis of its component amino-acids yields the amino-acid cysteine alkylated upon sulphur.

(3) Inhibition (inactivation) by the substrate analogue tosyl phenylalanine chloroketone, TPCK (6.6.1) is pH dependent and the pH profile implies that a group ionising at pK_a 8.3 must be present as the anion.

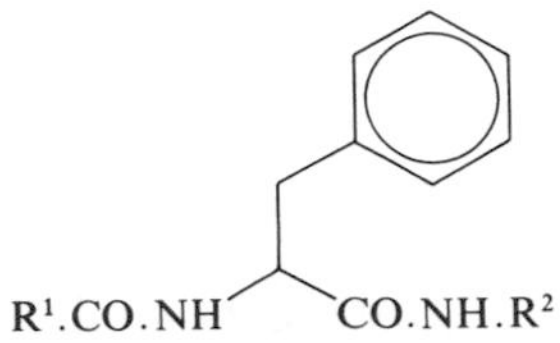

R^1.CO.NH — CO.NH.R^2

Substrate

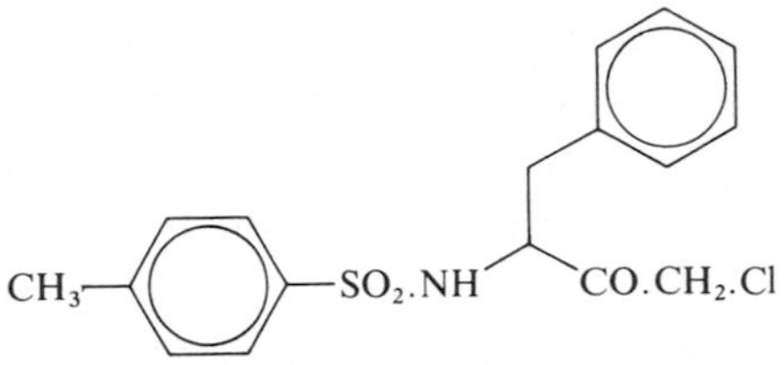

CH$_3$— —SO$_2$.NH — CO.CH$_2$.Cl

Inhibitor TPCK

(6.6.1)

Figure 6.6

Together, these results point convincingly to the involvement of the side-chain of cysteine in catalysis because our general chemical experience tells us that sulphydryl groups such as that present in cysteine are (1) readily attacked by mercuric compounds, and (2), (3) are readily alkylated by alkyl iodides and chlorides. The possibility that the first two reactions simply bring about some drastic alteration in the conformation of papain by preventing the formation of the active site is made more remote by the third observation. Here the inhibitor is closely structurally related to the substrate and we would therefore expect it to react at the active site of the enzyme just like the substrate. Again cysteine is alkylated and we obtain also the valuable physical evidence that a group of pK_a 8.3 is involved. This value is typical of a cysteine thiol. With this background knowledge, the chemist can set to work and study the mechanism of the reactions using synthetic, conceptual or mathematical models as appropriate (ref. 7).

Spectroscopic techniques in the study of enzymes. At first students of enzymes and of the reactions that they catalyse had to work with relatively crude tools, principally spectroscopic, inhibition and kinetic studies, but their conclusions on the nature of the active sites of many enzymes have been borne out by more penetrating modern probes such as electron paramagnetic resonance and nuclear magnetic resonance spectroscopy which observe environment-sensitive transitions of the spins of electrons or nuclei (see section 6.5 for examples). These techniques rely for the interpretation of their results on comparison with model compounds, an aspect of modelling which has been considered in the previous chapter. Along with kinetic studies, the technique which has shed most light upon the catalytic workings of enzymes has been X-ray crystallography (ref. 8). This technique, although lengthy and difficult, especially when applied to biological macromolecules which have limited stability to X-radiation and which are often difficult to crystallise, has a precise elegance that provides a direct link between the model of known chemical structure and the protein prototype. The characterisation of the components and structure of both the binding and the active sites affords an insight into the nature of the catalytic process. However, one diffraction experiment on one crystal of enzyme or derivative illustrates only one stage in the reaction. To build up a dynamic picture of catalysis, either many derivatives of the enzyme and substrate and, if possible, reaction intermediates must all be subjected to X-ray analysis, or one must construct a molecular model (see chapter 5) of the enzyme based upon the available coordinates of atoms from a single X-ray diffraction experiment. By manipulating, in structural models, the interacting substrate and amino-acid side-chains of the enzyme, it is often possible to come to an understanding of the catalytic process which is consistent with all known spectral and kinetic data. In this way, the most refined expression of the mechanism of action of chymotrypsin has been reached (ref. 9) (see section 6.3). However even when such a success has been achieved, the whole chemical story has not been told because the information to hand refers only to one case, that which evolution found to be most efficient. The call to model is again sounded.

6.1.3 *Some characteristics of models of biological systems*

Modelling recognises the strains upon other experimental approaches which complex molecules such as enzymes and proteins enforce and attempts to identify the important features of a biological system and to replicate them in a way which promises meaningful results. The success or failure of a model system in explaining the behaviour of its prototype depends clearly upon the precision with which these features (or parameters) can be identified and, most importantly, related from model to prototype. In this respect we must re-emphasise that it is not only the transformation from model to prototype which matters but, since enzymes carry out their catalytic business in a controlled micro-environment, the relationship between prototype and model environments also assumes importance. The great uncertainty of the transformations inherent in models of biological systems makes it a simple matter to set up as a model a system which performs the same reactions as the enzyme. Such a model is, however, valid only within the limits of certainty of the information upon which it is based. Like any scientific theory, a model should not be sterile but should have predictive power and when predictions drawn from the model turn out to be false, there must have been a conceptual error of design in terms of either environmental correspondence or parameter definition. As will be seen in section 6.7, even if the model is very good at explaining the general chemistry of its biological prototype, over-zealous extrapolation can lead to conclusions which are seriously in error. In cases where the transformations are vague or ill-defined, the danger exists that the essential features of catalysis by the enzyme may be simplified out of the model (ref. 10). Therefore as always, the problem for the modeller, given the background information, is to identify the important parameters of a complex system and to combine them into a model in as controlled a manner as possible. In attempting this often subjective restructuring it is important to consider the system in relation to the wider problems and principles of the field and to ask questions as one must such as – 'Is the biosynthetic scheme proposed a chemically reasonable process?' or, 'What is it that makes the enzyme catalysed reaction so astonishingly efficient and selective under such mild reaction conditions?' In the survey of examples which follows, we shall see how far the collective efforts of modellers have been successful in providing a basis for answering these questions. The range of chemistry covered is large, just as enzymic catalysis employs reactions characteristic of many different fields of chemistry from inorganic coordination compounds to ion transfer reactions. The examples reflect not only the approach of modelling, but also the wide expanse of chemical thought applied to one basic problem, the chemical nature of life.

6.2 Modelling and biosynthesis
6.2.1 *Biogenetic-type synthesis*

The success of the chemical approach to the elucidation of the many reaction pathways by which the molecules of nature are synthesised has, more than anything else, stretched the outlook of chemists to the biological horizon. The initial impulse was generated largely by one man, Robinson, whose perceptive intuition recognised structural similarities between certain alkaloids that indicated that they had a common natural precursor (ref. 3). It was not possible at that time to test the theoretical biosynthetic pathways by experiments such as the tracer labelling method described above, but nevertheless, Robinson demonstrated the feasibility of his biosynthetic concept in a laboratory synthesis of the fundamental structural unit of the tropine alkaloids, *tropinone* (figure 6.7) from a combination of molecules which very probably have close natural counterparts (ref. 11). This was the prototype *biogenetic-type synthesis.*

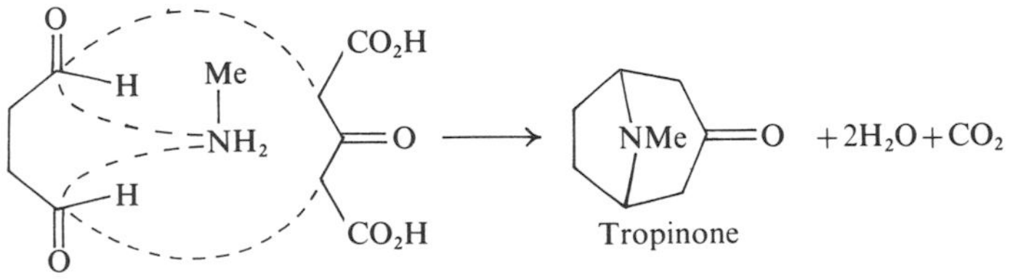

Figure 6.7. Robinson's biogenetic-type synthesis of tropinone.

After 1945, tracers aided the establishment of more complete biosynthetic pathways making available a prototype for modelling studies which followed two distinct paths.

Firstly, following Robinson, syntheses of complex natural products were conceived using a reaction sequence which was based upon an established biosynthetic pathway or upon a reasonable biosynthetic speculation. The chemist here accepts a challenge from nature to find a laboratory method that runs parallel to a natural reaction.

Biogenetic-type syntheses as models. As we have already seen, the correspondence of biological reactions to known chemical processes is well established and constitutes perhaps the firmest model–prototype relationship in the field of biological organic chemistry. The detailed philosophy of these biogenetic-type syntheses has been reviewed by van Tamelen (ref. 12). In such work, there is no attempt to mimic nature directly by carrying out reactions under physiological conditions of temperature, pH and ionic strength, for example, as required by the enzymes catalysing the natural reactions; the correspondence need not even be as close as substituting sodium borohydride in the laboratory for the natural reducing agent, nicotinamide

adenine dinucleotide (NADH see section 6.7), but the structural units
are combined in a manner parallel to nature, that is the prototype for
the model biogenetic-type synthesis is simply the biosynthetic sequence
looked at in its broadest terms. Commonly a synthesis follows its
prototype closely only at stages that are crucial to the construction
of a complex fundamental structural unit. Even under this limitation,
the yield of products is often low because under the *in vitro* conditions
the control over a number of possible reaction paths which an enzyme
exerts is absent. However, if the 'substrate' itself has particularly
severe steric requirements for reaction or is constrained to react in only
one way, for example, by use of appropriate protecting groups, novel
synthetic procedures can be established.

Maritidine – an alkaloid. A biogenetic-type synthesis of the alkaloid
maritidine (figure 6.8) illustrates these points. The prototype is the
well-established involvement of the oxidative coupling of phenolate
radicals (e.g. 6.8.3) in the biosynthesis of several classes of alkaloid
(ref. 13). The synthesis (ref. 14) is based upon the trifluoroacetylbis-
benzylamine (6.8.1) which is a very close model of the probable
natural intermediate (6.8.2). The possible coupling reactions of the
phenolate radicals derived from the synthetic precursor (6.8.1) are
limited by the protection of the oxidation sensitive benzylamine as a
very electron deficient trifluoroacetamide. This raised the yield of the
key intermediate, the dienone trifluoroacetate (6.8.4) (with three of the
four rings of the natural product already incorporated) to twenty times
that of any other reported model oxidative coupling. The synthesis of
maritidine was completed by conventional transformations, rather than
by further stages of modelling.

Isoprenoid compounds. In the foregoing example, control within the
model was established by a deliberate structural alteration. In contrast, *squalene-2,3-oxide* (6.9.2), which has been shown to be a substrate
for many enzymic systems which synthesise steroids and triterpenes
(ref. 15), has very precise intrinsic steric requirements because of the
folding of the C-30 carbon chain about itself. That the conformation
of this molecule could itself dictate the stereochemistry of cyclisation
to steroids or triterpenes by stereoelectronic effects was first hypo-
thesised in 1955 (ref. 16) and recently, the question has been examined
by biogenetic model studies. The important structural feature of squalene and its oxide is that the *trans* double bonds can find themselves

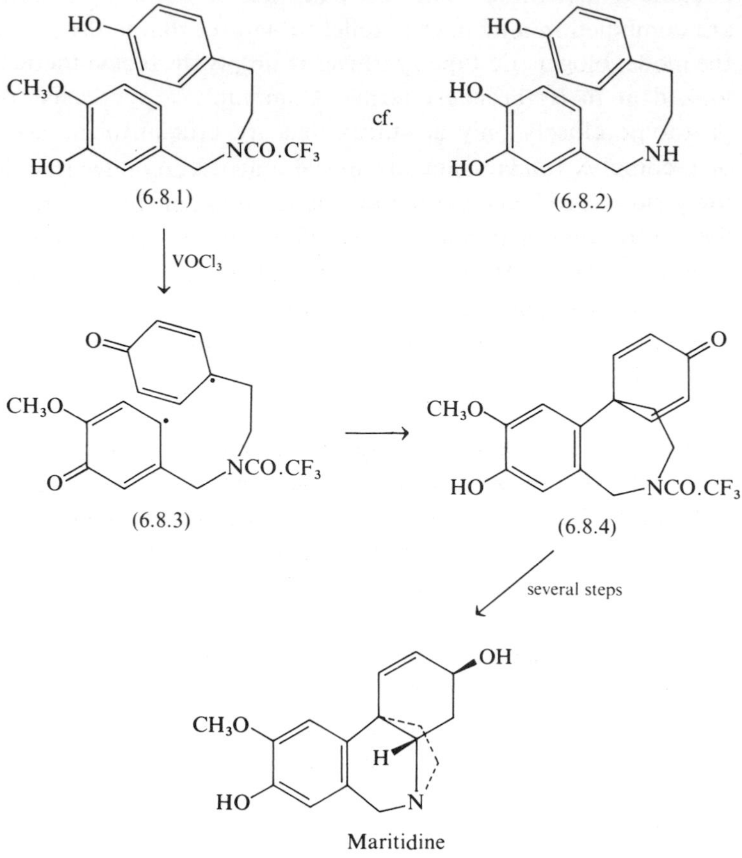

Figure 6.8. A biogenetic-type synthesis of (±)-maritidine.

in close proximity, as if coiled over one another, so that once a cyclisation reaction is initiated at one end, it should continue and produce the same stereochemistry regardless of the mode of initiation. Chemically, a carbenium ion mechanism of cyclisation is most favourable, and early model studies used another precursor for a carbenium ion, an acetal (6.9.3) (ref. 17). A simplified *trans* olefin was also used, but the products showed the natural *trans* fusion of the six-membered rings (6.9.4). A spectacular example of the 'self-control' of squalene oxide cyclisation is the *in vitro* production of the natural terpenoid ±-*malabaricanediol* from the dihydroxy squalene-2,3-oxide (6.9.5 and 6.9.6) correct to 9 asymmetric centres (ref. 18). A recent and potentially useful application of such cyclisations is Johnson's elegant synthesis of progesterone and also of estrone. The key step in the synthesis is a cyclisation of 3 of the 4 rings in a substrate (6.9.7) which differs from squalene-2,3-oxide in that one ring is preformed and one double bond replaced by an acetylene (ref. 19). The problem of environment correspondence is not of importance in these cases since essentially the squalene provides one and the same reaction environment both in model and natural cases.

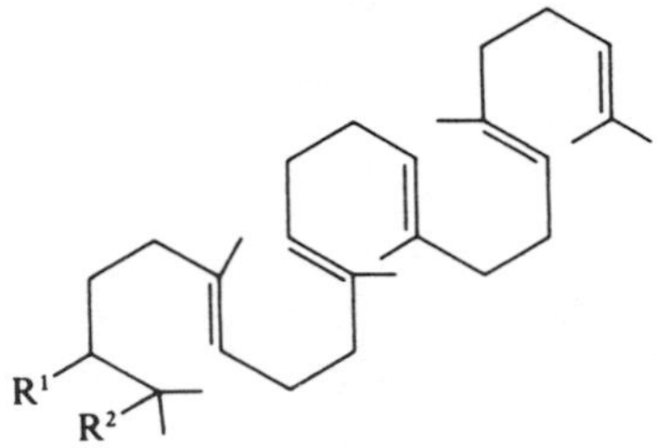

(6.9.1) $R^1 =$ H squalene

(6.9.2) $R^1 = R^2 = $ ⌇O⌇ squalene–2,3–oxide

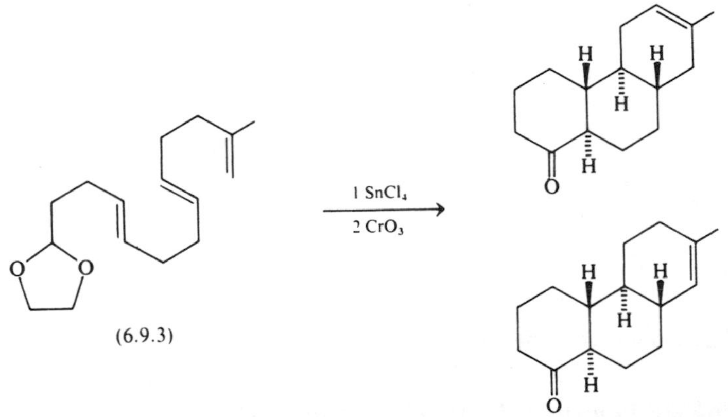

(6.9.3)

1 SnCl$_4$
2 CrO$_3$

50 % mixture of isomers

(6.9.4)

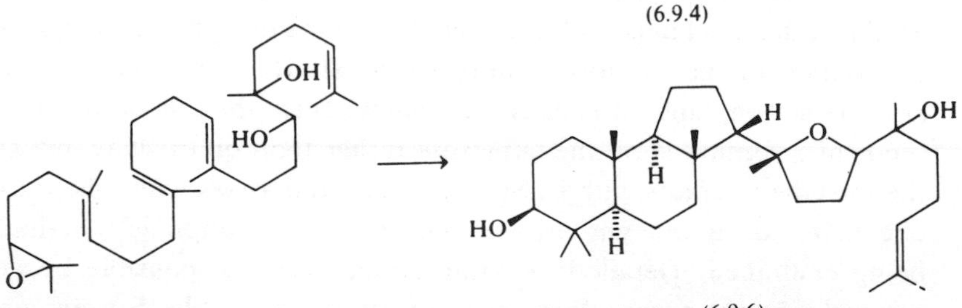

(6.9.5)

(6.9.6)

Figure 6.9. Model biogenetic cyclisations of olefins to terpenoids.

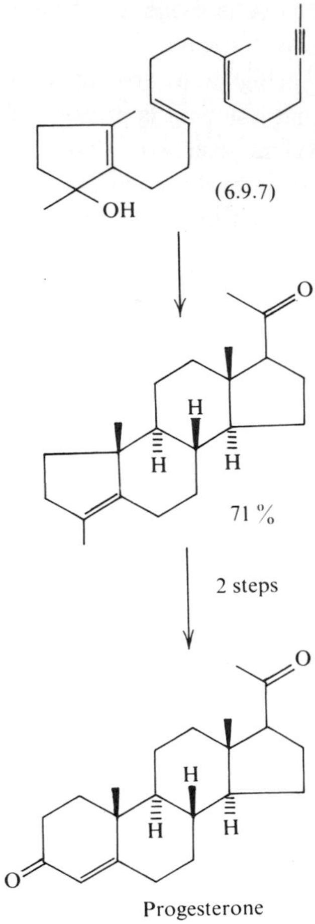

Figure 6.9 (*cont.*) Biogenetic-type steroid synthesis.

6.2.2 *The mechanism of biosynthetic reactions*

The second aspect of biosynthetic modelling is the attempt to establish the mechanism of each reaction in the biosynthetic sequence. Of course this has a close affinity to the study of the mechanism of enzyme action, since it is enzymes that perform the reactions, but the focus of attention is on the substrate rather than on the way in which the enzyme interacts with it. Such a case often arises when the stereo-chemistry of a biosynthetic condensation or coupling reaction is being examined. Detailed enzymic studies are not possible because purified enzyme preparations are not usually available. For example, *squalene* (6.9.1) is formed in nature by the reductive coupling of two farnesyl pyrophosphate units in the presence of nicotinamide adenine dinucleotide phosphate (NADPH). The stereochemistry of this important reaction has been shown by Cornforth (ref. 20) to involve inversion of configuration at one of the terminal methylene groups of farnesyl pyrophosphate (figure 6.10) and he postulated that one way

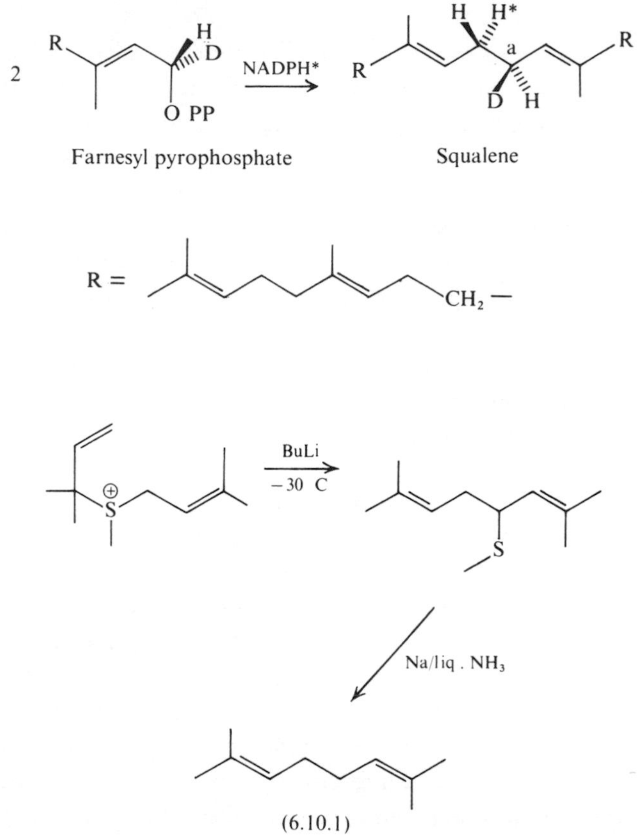

Farnesyl pyrophosphate   Squalene

(6.10.1)

Figure 6.10. The biosynthesis of squalene, and its model.

in which such a process might occur is via an intermediate sulphonium salt formed by interaction of two molecules of farnesyl pyrophosphate and a free sulphydryl group at the enzyme's active site. Several teams were quick to take this cue to examine the proposed novel transformation with laboratory analogues (ref. 21). All successfully showed that the rearrangement of a sulphur containing model of the proposed enzyme-bound intermediate did indeed give rise to a squalene-like compound on treatment with base (6.10.1). Unfortunately, the prototype for these models was constructed on the basis of a hypothesis which further work proved to be irrelevant to the biosynthetic pathway by the isolation and incorporation of C-30 precursors of squalene which are unlikely to be formed through a sulphonium salt intermediate (ref. 22).

6.3 Model studies of hydrolytic enzymes (without metal cofactors)

The enzyme which has been by far the most thoroughly studied, both by direct and by modelling methods, is *chymotrypsin* (E.C. 3.4.4.5), a proteinase which catalyses the hydrolysis of peptide bonds of proteins. In nature, it has a specific substrate – it will only hydrolyse peptide bonds between the carbonyl end of an aromatic L-amino-acid (phenylalanine, tyrosine or tryptophan) and another amino-acid, but it can be pressed to accept and hydrolyse many other substrates *in vitro*, albeit at a greatly reduced rate. This is indeed fortunate because without the use of such unnatural model substrates, which include both amides and esters, knowledge of catalysis and chymotrypsin would be very much less advanced.

6.3.1 *Kinetic studies with model substrates*

As we have seen already in the case of maritidine synthesis, one important possibility made available by a model compound is the control of reactivity, which arises by virtue of the relative simplicity of the model. The first studies aimed at establishing the mechanism of action of chymotrypsin took advantage of the tolerance of the enzyme for unnatural substrates and employed p-nitrophenyl esters, especially the acetate, as substrate. An important feature of modelling is illustrated here. When we have considerable freedom in choosing our model substrate it is sensible to select a molecule which will make the reaction easy to follow. Hydrolysis of p-nitrophenyl esters produces the yellow p-nitrophenate ion which permits easy monitoring of the reaction by spectrophotometry.

Control of enzyme catalysed reactions can be exercised by three modelling operations – by modifying the enzyme, or the substrate, or both. The most common course is to select a substrate like the p-nitrophenyl derivative which gives convenience in handling; this procedure avoids interference with the enzyme itself and the model–prototype correspondence is at its closest. Alternatively, the enzyme can be modified, for example by exchanging one metal ion for another that can be detected by spectroscopy. (This topic is discussed more fully in section 6.5.) Again, the model–prototype correspondence can be very close, if care is exercised in the choice of the modification. However, deductions from experiments which simultaneously vary both enzyme and substrate will have much less certainty, because the number of disturbed interactions will obviously increase, and will be difficult to define.

From the results of kinetic studies obtained using p-nitrophenyl esters as substrates for chymotrypsin, it became apparent that hydrolysis occurs by a two step mechanism known as a double displacement (ref. 23a). This implies the existence of an intermediate with the substrate bound in some way to the enzyme; in simplified form, a molecular interpretation of the kinetic data is given in figure 6.11.

$$E+S \rightleftharpoons (ES)^1$$
Michaelis complex

$$(ES)^1 \rightleftharpoons (ES)^2 + P^1$$
Acyl   Alcohol or amine
enzyme   leaving group

$$(ES)^2 \rightleftharpoons (EP)^2$$

$$(EP)^2 \longrightarrow E + P^2$$
Carboxylic
acid

E = enzyme   P = product   S = substrate

Figure 6.11. The kinetic mechanisms of chymotrypsin catalysed hydrolysis.

Two enzyme–substrate intermediates are indicated in this scheme – the first, the so-called Michaelis complex, is simply an association of the enzyme and the substrate without any chemical bond breaking or forming – the second is a species in which the acyl group of the substrate has been transferred to the enzyme to form a covalently bound acyl enzyme. The detailed course of a chymotrypsin catalysed hydrolysis is determined by the relative rates of the reactions represented in figure 6.11. If we use the freedom available to us as modellers and wish to detect the presence of the acylated enzyme intermediate, then we should arrange for the acylation step to be very fast. This has been done by using the very labile cinnamoyl p-nitrophenyl esters as substrate; the presence of the acyl enzyme was revealed by the ultra-violet spectrum (ref. 24) and it was suggested that the amino-acid side-chain which was acylated was probably the hydroxyl side-chain of serine.

Studies of the kinetic behaviour of hydrolysis at a variety of pH values showed that the rate of reaction was maximal at pH 7 implicating the involvement of an ionisable group of pK about 7 in catalysis. This result was most important because it provided the speculative basis for a vast number of model experiments. The most probable group ionising at pH 7 in a protein, assuming that the environment of the enzyme has not distorted things too much (a reasonable assumption for a hydrolytic process requiring that water be accessible) is the imidazole side-chain of histidine (ref. 25). If this limited assumption is accepted, a new problem can be posed, and one ideally suited to modelling, namely how can imidazole act as a catalytic species contributing to rates of hydrolysis as large as those of chymotrypsin? In the search for an answer the sophistication of models of this aspect of chymotrypsin chemistry has gradually been increased to include more parameters of likely importance in the enzyme catalysed reaction.

6.3.2 *Models relating to catalysis by imidazole – approaches to active site models*

The easiest first approach to the study of catalytic power of imidazole is to determine the kinetics of hydrolysis in the presence of imidazole (ref. 26). No especially fast rates of reaction were found, and instead of solving one problem, these experiments posed many more. It became clear, however, that more than the well-understood acid–base catalysis was involved, and that a new form of catalysis, nucleophilic catalysis, in which the catalytic nucleophile (imidazole here) forms a covalent bond with the substrate, was also playing a part (refs. 27, 28a). This is illustrated by figure 6.12 (ref. 26).

As is common with most pioneering research efforts, the spin off from these fundamental studies using simple, well-defined models, has been as important in the general understanding of catalysis in organic reactions as in the elucidation of the mechanism of the enzyme catalysed reaction (ref. 28b).

The first extension to include parameters of likely importance in the enzyme catalysed reaction was to bring the reactants into proximity by localising them in the same molecule, as if they were bound close

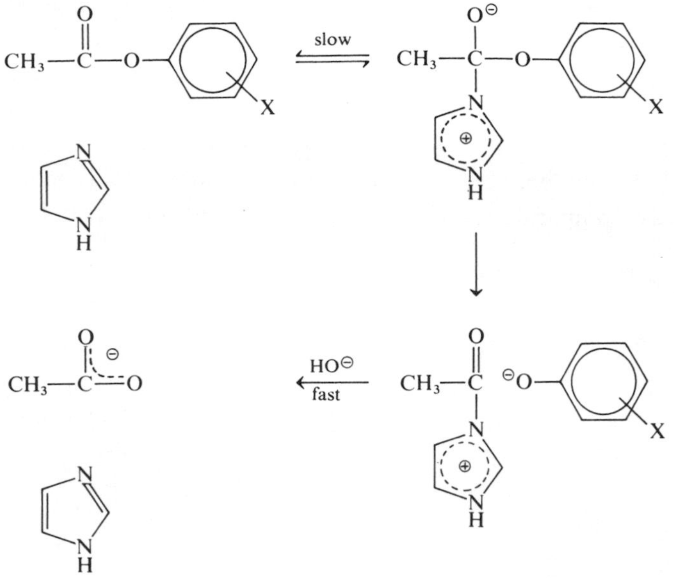

Figure 6.12. A mechanism of imadazole catalysed ester hydrolysis.

together on the active site of the enzyme. This arrangement is illustrated by the imidazoyl butanoate ester (6.13.1). A large rate enhancement of hydrolysis by the model nucleophile (imidazole ring again) relative to catalysis by free imidazole was observed (ref. 29). The rate constant is very similar to that of the acetylation of chymotrypsin (figure 6.11), but because of the absence of functional groups to carry out deacylation in the model, the overall rate of hydrolysis of the ester was much slower than when chymotrypsin was the catalyst. With this model, the importance of the spatial closeness of reacting groups in enzyme catalysed reactions has been demonstrated; 'approximation', as Jencks has dubbed this effect (ref. 28a), is one of the fundamental causes of the enormous catalytic efficiency of enzymes. A wealth of examples of proximity effects in chemical reactions have been discovered, and for a critical discussion of both theory and examples, the reader is referred to Jencks's excellent book.

Taking account of the fact that enzymes are large polypeptides, Sheehan attempted to model the active site of chymotrypsin with a relatively simple pentapeptide (6.13.2) containing both the presumed nucleophilic and acylating sites (histidine and serine respectively). Only a four-fold rate increase relative to free imidazole was observed (ref. 30). This turned out to be a poor model of the enzyme's active site and the result eloquently emphasises that the conformation of the active site of an enzyme is controlled not by the reactive amino-acid side-chains but rather by the concerted action of the whole protein. In other words, a model active site removed from the environmental control of the enzyme, if this were possible, would not show a reactivity comparable to the enzyme in its entirety. This is a further

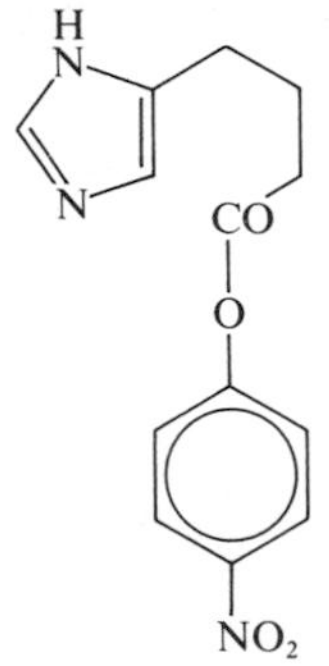

(6.13.1)

Thr.Ala.Ser.His.Asp

(6.13.2)

(6.13.3)

(6.13.4)

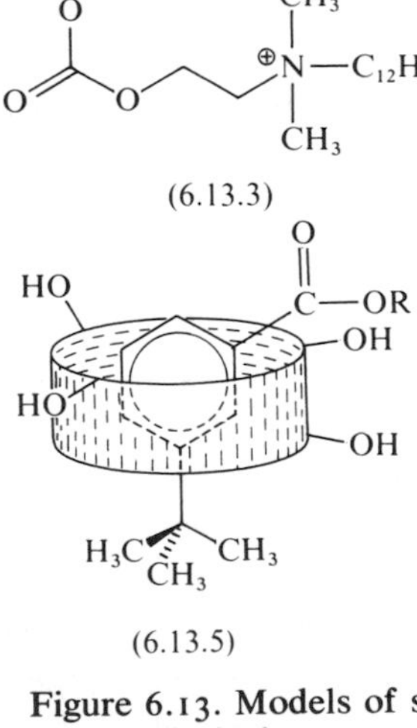

(6.13.5)

Figure 6.13. Models of several aspects of chymotrypsin catalysed ester hydrolysis.

example of the importance of environment interactions in both model
and prototype in enzymic systems.

6.3.3 *Substrate binding and model studies*

An interaction of great importance in the interior regions of
enzymes, which are protected by the aqueous environment of the cell,
is the mutual attraction of hydrophobic groups (ref. 28c, 31). This has
obvious relevance for the specificity of chymotrypsin for peptide bonds
adjacent to hydrophobic aromatic amino-acid side-chains and has been
incorporated into two model systems. In molecules (6.13.3) and
(6.13.4), the mutual attraction of the long fatty chains provides a
binding interaction between the model substrate (6.13.3) and catalyst
(6.13.4). This binding can be detected in the kinetics of the hydrolysis
which reveal the presence of a substrate–catalyst complex that is
kinetically directly analogous to the Michaelis complex of enzyme
catalysed hydrolysis (figure 6.11), a result that confirms in a model
the nature of interaction within a Michaelis complex (ref. 32). Further
background information relating to hydrophobic interactions has been
obtained from the study of micelles of molecules containing both a
hydrophobic and a hydrophilic region (ref. 33). Very high rates of
hydrolysis have been found. Such systems provide an opportunity to
study proximity effects in a weakly interacting system that is a model
for the interactions between enzyme and substrate that control
specificity.

The models we have just examined involve little abstraction from
the presumed reactive features of chymotrypsin. In contrast, Bender
has attempted to combine the catalytic properties of an enzyme in a
model that is synthetic both in conception and preparation (ref. 34).
Cyclic polydextrans (polysaccharides) known as cycloamyloses
(6.13.5) form a cylindrical tube into which substrate molecules can
penetrate and bind. Substituted phenyl acetates, for example, bind into
this cavity in different orientations dictated by the size and position of
the substituent. Thus in *ortho* and *para* substituted t-butyl phenyl
acetates, the ester group (substrate) is forced away from the hydroxy
groups protruding from the edge of the cylinder which form the model
active site. On the other hand, the *meta*-substituted derivative (6.13.5)
binds in an orientation which brings the ester linkage close to the active
site and accordingly, the highest rate of hydrolysis is found in this case.
One could say that the model has a high specificity for aromatic esters
bearing bulky *meta* substituents. Generalising from these results,

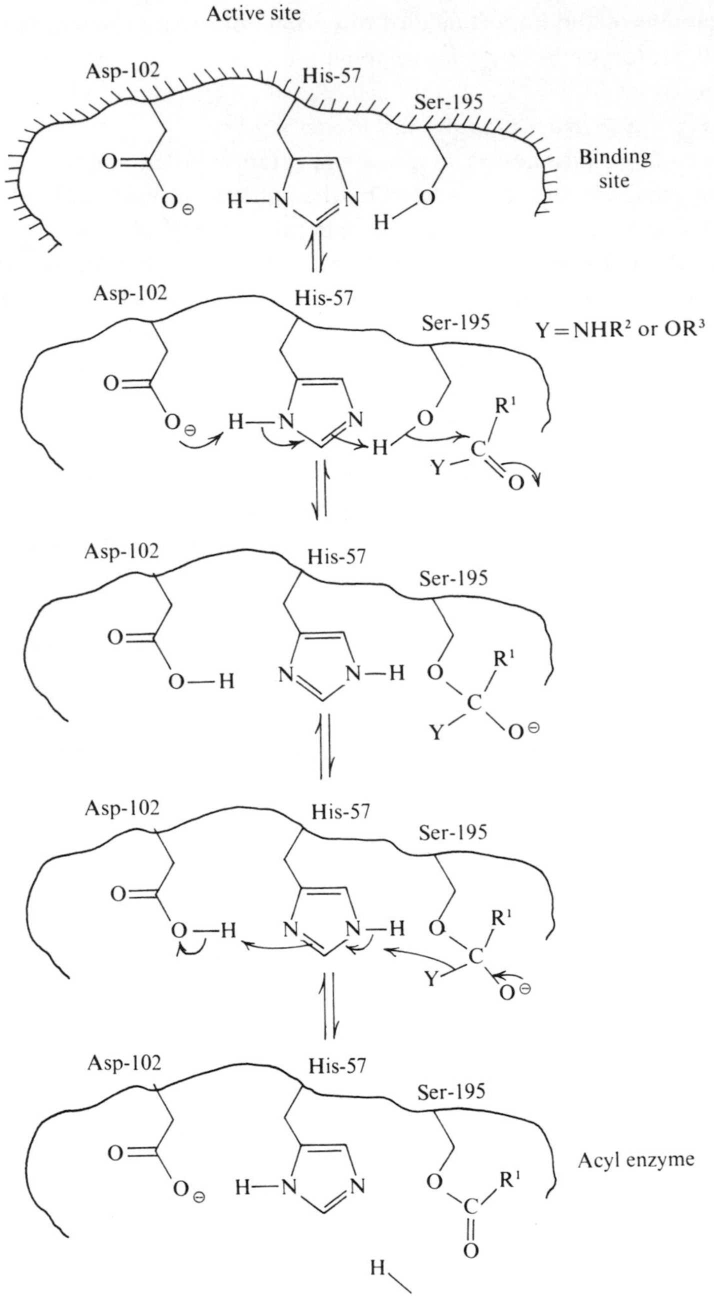

Figure 6.14. The acylation of chymotrypsin – a mechanism deduced from kinetic and X-ray crystallographic studies (adapted from Gray, ref. 7).

Bender emphasised that the geometric coercion of substrate into the active site, thereby bringing the reactants into optimal relative positions for catalysis, is a very powerful means by which enzymes can exert specificity (ref. 35). The rates of hydrolysis catalysed by chymotrypsin are those of L-amino-acids having an aromatic side-chain (ref. 23b), and the source of this specificity has recently been traced by X-ray crystallography to a non-polar pocket adjacent to the active site, which can only accommodate a hydrophobic side-chain of the L-absolute configuration (ref. 36). Bender's extrapolation from model results thus finds confirmation in the most closely related prototype. This is an excellent example of how modelling can help to indicate the relative importance of various parameters in a prototype.

6.3.4 *Models related to peptidases assessed*

These modelling procedures can be applied generally to many enzyme catalysed reactions. For example, models of the cysteine proteinases (relatives of chymotrypsin, that contain a sulphydryl function instead of a hydroxyl group in the active site) have been constructed in ways similar to those we have considered (ref. 37). All stages of mechanistic investigations into chymotryptic hydrolysis have found substrate and enzyme models helpful, although model studies alone, of course, cannot elucidate the mechanism. The importance lies, as we have seen, not only in the understanding of enzyme catalysed hydrolysis but also in a more general understanding of catalytic processes in organic reactions (ref. 28b). One wonders whether such valuable effort would have been expended on the study of imidazole as a catalyst had it been realised earlier that the nucleophilic group involved in the acylation step of chymotrypsin catalysed hydrolysis is not the imidazole side-chain of histidine but the hydroxyl group of serine polarised into a highly nucleophilic state by the histidine residue, which itself is involved in proton transfer with a neighbouring aspartic acid, an environmental effect (ref. 9) (figure 6.14).

6.4 Model studies of haemoproteins

So far, we have considered enzymes that form their active site from a grouping of amino-acid side-chains, but this is not the only way in which an active region can be composed. In many proteins, especially in those which conduct redox processes as part of their biological function, a non-protein molecule bound into the protein by electrostatic, non-polar or covalent interactions is an essential component of the active site. An important class of such proteins are the *haemoproteins* which contain a porphyrin iron complex or *haem* (6.15.1) as the non-protein component or *prosthetic group*. The function of the haemoproteins is to carry out sequential oxidation processes in the cell, in which the final oxidant is molecular oxygen carried through the blood by the haemoprotein *haemoglobin*. Because they do not catalyse any chemical reaction but are essentially energy carriers, some haemoproteins cannot strictly be called enzymes. However, the behaviour of the protein component shows many similarities with that of enzymes in the nature of the interactions occurring within it. Two groups of haemoproteins may be distinguished – the oxygen carrying haemoproteins *haemoglobin* and *myoglobin* and the electron transporting *cytochromes*. The latter are involved both in electron transport in animals and also in plants during photosynthesis. In every haemoprotein, the reactive site is a porphyrin-iron complex which is differently situated in each protein and may also differ in the structure of the porphyrin side-chains.

6.4.1 *Haemoglobin and myoglobin*

The application of models to these proteins has followed a different course from that in the case of chymotrypsin and its relatives, because the structure of myoglobin was known (it was the first protein whose structure was successfully determined by X-ray crystallography) (ref. 38) and this provided much detailed background information which greatly aids modelling. Although the functions of the haemoproteins in nature have long been known in detail (refs. 39 and 40), many aspects of the mechanism of their reactions have remained obscure, largely because the effect of the environment on the properties of a metal ion is not understood. Not only the ligands, but also other nearby groups are important factors in the environment. The redox potentials of the cytochromes, for example, are largely controlled by the environment of the haem and the elucidation of the factors that regulate the properties of the metal ion remains the prime problem to be solved in haemoprotein chemistry. Two problems of particular import for haemoglobin are much better understood as a direct result of model experiments namely (1) the nature of the bonding of oxygen to the haem iron, and (2) the remarkable stability of the reactive iron(II) oxidation state in the presence of molecular oxygen.

Spectroscopic properties of the prototype. Because of synthetic complexity it is difficult to construct the sort of versatile model of haemoproteins that would be useful in investigating environmental factors and accordingly, the first compounds to be examined as models were the readily available porphyrin complexes of transition metals in various oxidation states. Porphyrins and their metal complexes have highly characteristic visible spectra (ref. 41); almost every model system has used this property for comparison with the prototype haemoproteins. For example, it is easy to prepare derivatives of haems by increasing the coordination number of the iron from 4 in the haem to six, usually with basic ligands such as chloride, pyridine, imidazole or even water. A close correspondence in spectra was found between haemoglobin and the *imidazole haemochrome* (6.15.2) from which it was suggested that one of the ligands of iron in haemoglobin should be the imidazole side-chain of histidine. We shall discuss a more sophisticated model which takes account of this result later in this section. An immense number of variations on this simple spectroscopic correspondence theme have been studied but it has not been possible to generalise from these results and describe the nature of environmental effects. This is hardly surprising since these models do not take account of environment nor of the nature of oxygen bonding to iron. Let us examine some models relevant to the latter problem first.

Synthetic oxygen-carrying compounds. The credit for the discovery of a whole class of models of haemoglobin must go to the United States Navy rather than to a biologically oriented inorganic chemist. During World War 2, the U.S. Navy was looking for something cheaper, lighter and less bulky to supply oxygen to their submarine crews than the conventional compressed air, and someone hit upon the idea of preparing chelated metal derivatives which would store oxygen by reversible binding of oxygen to the metal in analogy with haemoglobin. In the event, cobalt derivatives proved more satisfactory than iron, but even so, these novel oxygen sources did not attain sufficient cost effectiveness to be used in combat submarines. Their chemical legacy has fortunately been much greater than their military efficacy.

Perhaps the most studied chelates have been *cobalt salicylidene ethylenediamine* derivatives (6.16.1) which will undergo several cycles of oxygenation and deoxygenation without deterioration of efficiency. Cobalt compounds were found to be less susceptible to oxidative deterioration (usually to bridged μ-peroxy compounds (6.16.2) (ref.

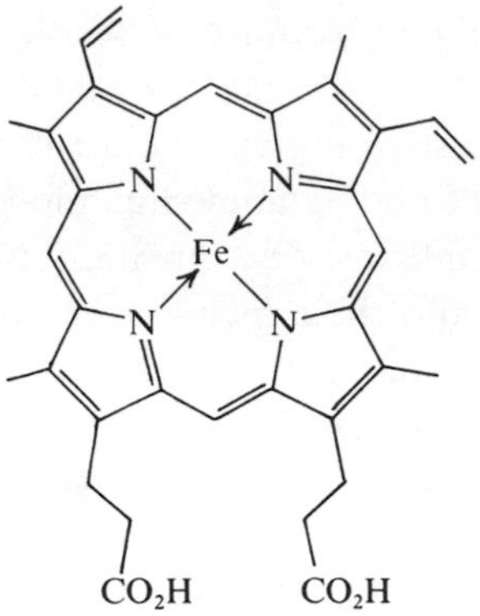

(6.15.1)

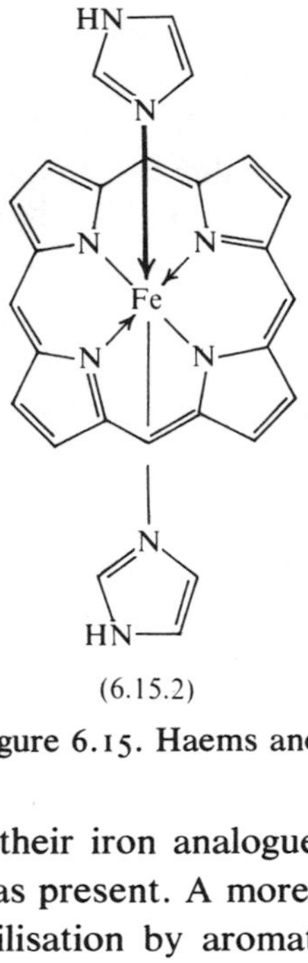

(6.15.2)

Figure 6.15. Haems and haemochromes.

42)) than their iron analogues except when some unusual stabilising feature was present. A more recent, highly ingenious model provides such stabilisation by aromatisation of the cycloheptatrienyl ring of the ligand to tropylium on uptake of oxygen (6.16.3) (ref. 43). They, and many other complexes, have common features which define the chemical requirements for a molecule which will bind oxygen reversibly. The most important to emerge from these models are that (1) the metal ion must be firmly coordinated such that a binding position remains vacant for oxygen – this implies that the fundamental ligand should generally be a square planar tetradentate ligand for first

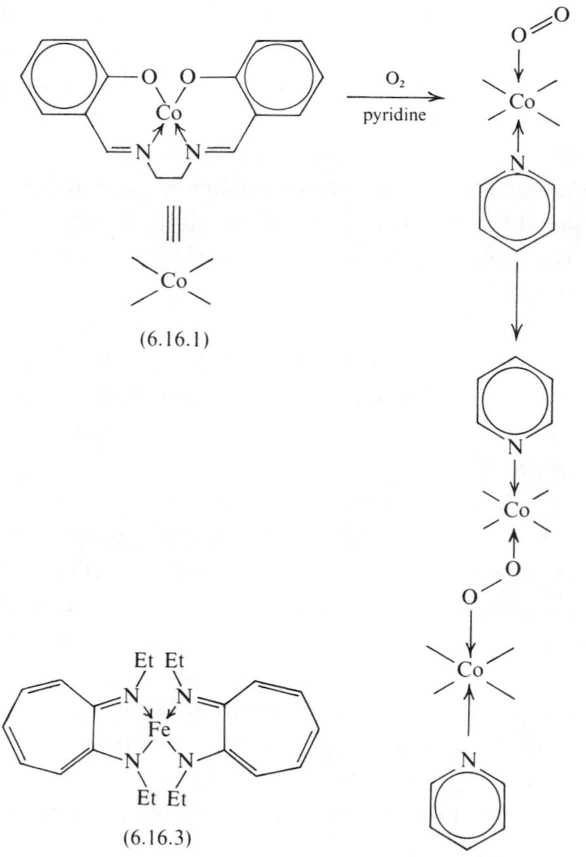

Figure 6.16. Some oxygen-carrying transition metal chelates.

row transition metals – and (2) the formation of peroxide species must be inhibited. These requirements, although established by work on models which are distant structural relatives of the prototype haemoproteins, do seem applicable to haemoproteins also.

It is worth noting that a wide range of models in which the prototype–model transformation shows simplification down to only one or two features of correspondence, as in these oxygen binding chelates, can provide as much definitive information as a single model that possesses many corresponding features. In the former case, the accruing bonus of chemical knowledge is larger. A somewhat similar situation exists with many models of metalloenzymes discussed in section 6.5.

Correlation of models with prototype. Let us now relate these conditions to our prototype haemoglobin containing its iron chelated by a porphyrin. The porphyrin ring (6.15.1) is one of the most powerful chelating agents known (ref. 41) capable of extracting metal ions from chelation by other ligands such as acetylacetone. Thus the first criterion deduced from models correlates well with the prototype. How the second criterion is fulfilled in nature is best understood by reference to a more sophisticated model of haemoglobin which brings the environmental question into play.

An imaginative model of haemoprotein action. Before the X-ray structure determination of myoglobin was completed, Wang devised a model system which he hoped would shed some light upon the mode of coordination of oxygen to iron and also explain the remarkable stability of the ferrous oxidation state during oxygenation of haemoglobin and myoglobin (ref. 44). As his starting point he had only the chemical and spectroscopic evidence indicated earlier in this section. He attached particular importance to the suggestion that imidazole (again from histidine) was a probable ligand of the iron in addition to the porphyrin. In consequence he chose to embed a model haem (an iron complex of a porphyrin dicarboxylic acid diethyl ester) in a matrix of polystyrene containing the imidazole derivative 1-(2-phenylethyl)-imidazole – these two components model the protein both as a fundamental structural support and as a ligand at the presumed active site (figure 6.17).

The system was found to be capable of mimicking oxygenation of myoglobin, and spectra indicated that the iron remained ferrous throughout. Why does the model combine so well with oxygen? Clearly it has a free coordination position for oxygen because the iron is only pentacoordinate (four positions occupied by the porphyrin and one by the so-called proximal imidazole); it should be recalled that this is also a conclusion from the less sophisticated models discussed above. The presence of a second imidazole ring further away from the haem may be envisaged as a stabilising factor in the binding of oxygen. Secondly, why is the ferrous state stable in the presence of molecular oxygen? The most important structural feature of this model is that the haem is enveloped in an extremely hydrophobic environment, to wit the polymer matrix. This non-polar and non-acidic environment inhibits every possible mechanism by which iron(II) can be oxidised to iron(III) in the presence of oxygen. On completion of the X-ray study

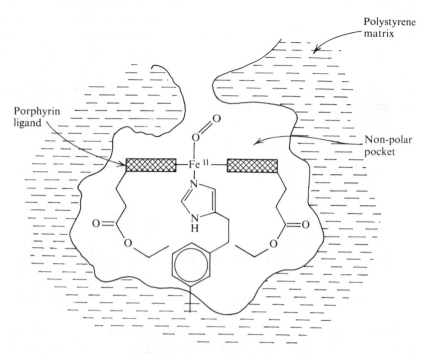

Figure 6.17. Wang's synthetic model of haemoglobin and myoglobin (ref. 44).

(ref. 38), it became apparent that a beautifully close correspondence between model and prototype existed. The haem was found to be located in a non-polar pocket in the centre of the protein and that iron was also coordinated by the imidazole group of a histidine residue with a second histidine a little further from the haem on the opposite side, just as in the model. Wang's mechanistic proposals have found acceptance as a description of these aspects of haemoglobin chemistry. The success of a model in which the environment of the active site was appropriately transformed speaks for itself.

6.4.2 *Cytochrome-c – an irrelevant model*

It has not always been Wang's good fortune to recognise the important parameters of a biological system so precisely. He observed that the bis-imidazole complex of iron(II) (figure 6.18) produced 1-phosphoimidazole on oxidation by air in the presence of phosphate (ref. 45), and consequently suggested that this might be a model for the coupled oxidative phosphorylation in which the cytochromes participate in nature (ref. 40). A radical mechanism such as that illustrated by figure 6.18 was proposed. On the other hand, Castro and Davis (ref. 46) showed that oxidation of iron by electron transfer need not involve any ligands other than the porphyrin itself. Their mechanism, based upon deuterium and tritium exchange experiments (as indicated in figure 6.19) implies that oxidation occurs initially at the periphery of the porphyrin ring.

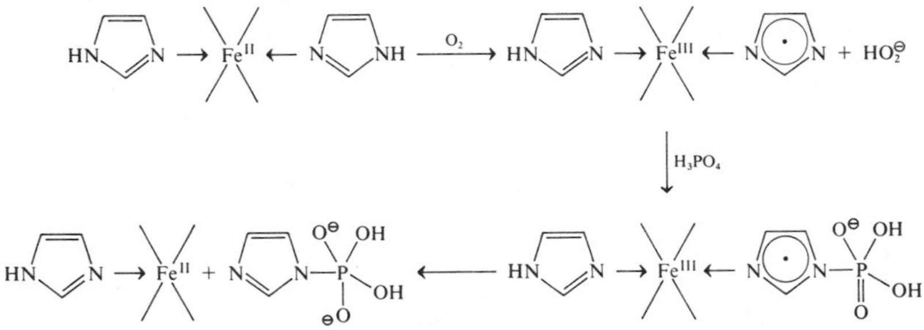

Figure 6.18. Wang's model reaction of oxidative phosphorylation.

The two models are clearly incompatible as representations of the biological reaction. At present, Castro and Davies appear to be nearer the truth, at least in the case of the best characterised cytochrome, cytochrome-*c*, which, according to a recently completed X-ray crystallographic study (ref. 47), has its haem very rigidly embedded into the protein with an octahedrally coordinated iron by the porphyrin, histidine and methionine. Most significantly, one edge of the haem protrudes into the surrounding medium. This result, although supporting the peripheral oxidation mechanism for cytochrome-*c*, by no means invalidates Wang's experiment because there is no reason for the other cytochromes with their markedly different redox properties to react by the same mechanism.

We have seen how models have contributed greatly to solving the riddles of haemoprotein chemistry, but the present situation is still unsatisfactory because an understanding of the very basic problem of environmental effects upon the coordination and redox properties of metal ions remains unattained. This difficulty crops up again in models of metalloenzymes (section 6.5). Studies aimed at the environmental problem are amongst the most open and important contributions that modelling techniques can make to an important field of biochemistry in the future.

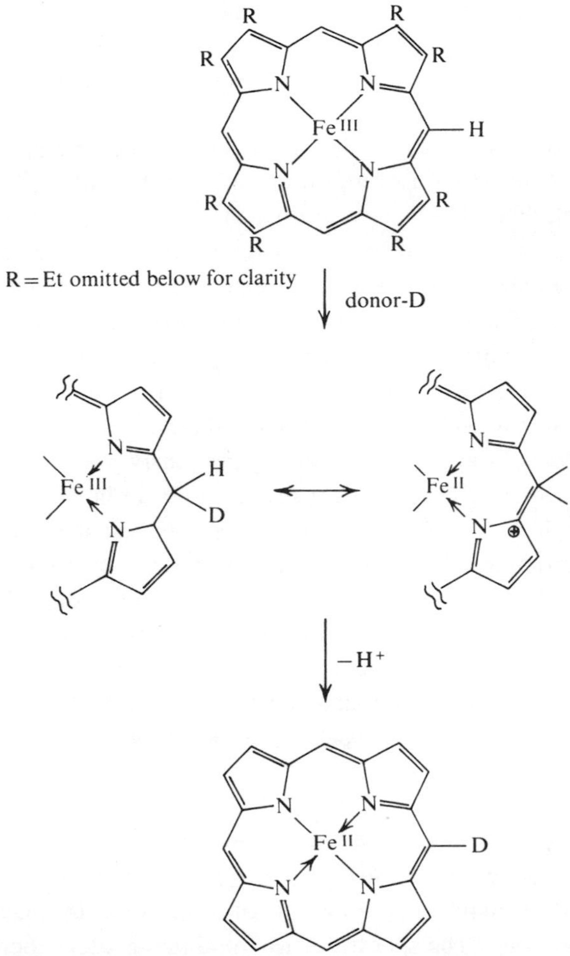
R
R
R
R
N
N
Fe III
H
N
N
R
R
R
R
R = Et omitted below for clarity
donor-D
N
H
Fe III
D
N
N
H
Fe II
D
N
⊕
−H⁺
N
N
Fe II
D
N
N

Figure 6.19. Castro and Davies' model reaction (here shown in the reductive sense).

6.5 Model studies of enzymes containing functional metal ions

The merging of biochemistry and inorganic chemistry begun with the haemoproteins has developed further into the study of enzymes which have a metal ion involved in the catalytic process, that is to say bonded directly to the protein rather than combined in a prosthetic group. Two main groups of these metalloenzymes may be distinguished – the *proteolytic* or *protein hydrolysing metalloenzymes*, and the *oxidising metalloenzymes*. One metalloenzyme of a third type, carbonic anhydrase, will also be mentioned briefly. The major problem common to all these enzymes is, of course, the function of the metal, and the first question which must be answered in this connection is what ligands bind the metal ion and in what configuration? One obvious modelling approach is to examine the spectra of the metalloenzyme and of simple coordination compounds of metals and to attempt a correlation of the metal ion's environment in the enzyme, revealed by its spectra, with complexes of known structure. As in the haemoprotein models described above, the wider the range of models studied by both metal and ligand substitutions, the better the chance of general results emerging. In other words, the metal must, if possible, be varied in both model and enzyme and the various complexes must be examined by several spectroscopic techniques. Such a complete study is an indirect probe into the metal ion's environment, particularly into the geometry of the ligands around the ion, because any unusual behaviour within the enzyme will be readily spotted under the closely controlled condition of the comparison (chapter 5). In fact, as has recently been pointed out by Malström (ref. 48), it is very often found that the behaviour of a protein bound metal ion cannot be adequately described by model coordination compounds. Before examining some contrasting examples illustrative of the unusual ligand properties of enzymes, a passing comment on the use of spectroscopic probes in enzymes is in order.

6.5.1 *Spectroscopic probes and metalloenzymes*

The principle of all probe techniques is the insertion of a spectroscopically 'visible' component of a protein in place of a similar 'invisible' one. The spectra of the modified protein then reflect the environment into which the probe has been placed and the nature of this environment can in principle be deduced from comparison with the spectra of simpler known compounds. Clearly success is dependent upon the availability of a sufficient number of relevant simple model compounds to make the spectrum of the probe derivative interpretable.

Some of the metal ions found in biological systems are coloured and thus serve as their own probes for visible spectroscopy; enzymes containing iron, cobalt and copper can easily be studied in this way. However, not all metal ions are coloured and in an enzyme containing zinc, for example, visible spectra would be useless unless the zinc were replaced by a coloured ion of similar size and charge. Cobalt commonly serves this purpose. Alkali metal and alkaline earth ions, uncommon in enzymes, but most important in the balanced functioning of a cell, can be 'dyed' by exchange with a rare earth metal ion of similar size (e.g. Nd^{2+} for Ca^{2+}) which binds into the same environment at the natural ion. The very sharp lines of the resulting spectra are easily characterised (ref. 49). Perhaps more sensitive to environment than the electronic transitions in a metal complex which give rise to the visible and ultraviolet spectra are nuclear and electronic spins. Any paramagnetic metal ion has an electron paramagnetic resonance spectrum which is characteristic of the metal ion, its oxidation and spin states, and the nature of the coordination. Comparison of the epr spectra of models and enzymes has been particularly useful with the oxidising enzymes; the case of xanthine oxidase which contains molybdenum and iron is discussed later. A further extension of this technique employs organic compounds, usually paramagnetic nitroso derivatives, which can be specifically attached to functional groups in the protein (ref. 50). The resulting epr spectrum then reflects the new environment in which the nitroso group finds itself. Nuclear spins which are useful probes are very rare and only $^{205}Tl(I)$, which has the same spin quantum number as the proton, has been employed. It serves as a substitute for alkali metal ions (ref. 51).

6.5.2 *Carboxypeptidase* – '*model enzymes*'

The first example of a metalloenzyme, carboxypeptidase, reflects the difficulty in obtaining reliable model–prototype correlations in this field. Carboxypeptidase (E.C. 3.4.2.1, 3.4.2.2) is a zinc-containing enzyme which catalyses the hydrolysis of proteins from the C-terminus (that is from the terminal amino-acid which has a free α-carboxylic acid group) as part of the digestive process. Many metal exchanges have been made with this enzyme and from the results of kinetic measurements on a series of such model *enzymes*, an order of reactivity was assembled for the dipositive metal ions catalysing peptide bond hydrolysis as follows:

$$Co > Zn > Ni, Mn \| Cd, Pb, Hg, Cu.$$

Although the cadmium enzyme could bind a substrate, along with all the others to the right of the vertical lines, it could not catalyse hydrolysis (ref. 52). Hydrolysis of a peptide bond can occur by either acid or base catalysis, and since we are dealing here with cations as catalysts, the mechanism of hydrolysis must be essentially (Lewis) acid catalysed. One would therefore expect that the most efficient peptide bond hydrolysing catalyst would be the strongest Lewis acid, or in other words, the best electron acceptor (ref. 53). This postulate was consistent with studies on various complexes containing sulphur and nitrogen ligands such as cysteine ligands presumed to be involved at the catalytic site on the basis of inhibition experiments (ref. 52). However, both the postulate and the model complexes lead to the conclusion that, of the metal ions to the left of the slash, manganese should be the most reactive and cobalt the least, exactly the opposite of what is found. Clearly, then, some environmental effect operative within the enzyme is completely distorting the picture, or we are postulating an erroneous function for the metal ion. A further result from the substituted model enzymes, that all the metal ions will take part in the hydrolysis of esters, suggested that geometrical factors cannot account for the reversal of activity order. If, however, the function of the metal ion is mainly to bring the reactive groups into the optimum relative arrangement for hydrolysis, then the stability of the complex formed rather than the acidity of the ion will be the dominating parameter that controls the overall efficiency of the enzyme. This can be measured in terms of the free energy of binding ligands, such as those used in the earlier study, and the order of decreasing stability (binding free energy) was found to be:

$$Hg > Cu > Pb, Cd > Zn > Co > Ni > Fe > Mn.$$

This sequence now explains the catalytic efficiency of a metal ion. Mercury, on the one hand, binds sulphur ligands too strongly thus forming a very tightly bound complex which has insufficient lability to catalyse hydrolysis; it acts as an internal inhibitor by blocking the active site in a complex. On the other hand, ligand binding to manganese is too weak to bring the enzyme into the optimum conformation for catalysis. Nature therefore chooses a working compromise between these factors with zinc although it is surprising that cobalt, which forms the most active enzyme *in vitro*, was not selected by evolution.

What lessons for modelling can be learned from this lengthy tale? Perhaps most important is that if such a wide range of model derivatives had not been examined, the first postulate might have been considered an adequate expression of the mechanism of carboxypeptidase and the function of the metal ion incompletely interpreted. At least it was possible in this case to obtain a reasonable correlation with simple models, although not without difficulty.

6.5.3 *Carbonic anhydrase – novel coordination chemistry*

The enzyme carbonic anhydrase (E.C. 4.2.1.1) which is found in blood presents even greater problems. It is important in maintaining the acid–base balance in blood and, through the intermediacy of a zinc ion at the active site, catalyses the reaction

$$CO_2 + H_2O \xrightarrow{\text{carbonic anhydrase}} H^+ + HCO_3^-.$$

If cobalt is substituted for zinc, then the enzyme has an unusual visible spectrum which correlates well with neither typical tetrahedral cobalt complexes (e.g. $Co(NCS)_4^{2-}$) nor octahedral complexes (e.g. $Co(NH_3)_6^{2+}$) nor even the pentacoordinate trigonal bipyramidal derivatives (e.g. $Co.Br[N(CH_2.CH_2.N.Me_2)]_3^-$). The enzyme's spectrum is something unknown in models and is best described as between tetrahedral and trigonal bipyramidal configurations (ref. 54). It has again proved impossible to represent the environment of the metal ion by model compounds; the completion of an atomic resolution X-ray study on carbonic anhydrase may well be required before chemists even know what sort of complex they must make. By now, workers in this field are well aware of these problems and it appears that it is more the exception than the rule that coordination in metalloenzymes is of similar geometry to that in the simple compounds which usually serve as models. Consequently, in the words of two leading researchers in this field, Williams and Vallee (ref. 55), 'future models should incorporate the emerging structures of metal complexes in enzyme centres which may deviate significantly from simpler systems currently known'.

6.5.4 *Oxidases – complex enzymes*

In contrast to the foregoing two examples, models of coordination of the molybdenum present in xanthine oxidase seem to be exceptions to Williams's and Vallee's generalisation. Xanthine oxidase (xanthine:O_2 oxidoreductase E.C. 1.2.3.2) is a multicomponent enzyme, containing not only molybdenum but also iron and a flavin, which together catalyse the oxidation of xanthine to uric acid (figure 6.20).

Epr spectra of the enzyme obtained with sophisticated rapid mixing and stopped flow techniques showed a characteristic group of signals assigned to the paramagnetic d^1 Mo(V) species (ref. 56). At physiological pH, the model complex (6.20.1) shows an epr spectrum identical in detail with the enzyme's spectrum (ref. 57). It can therefore be

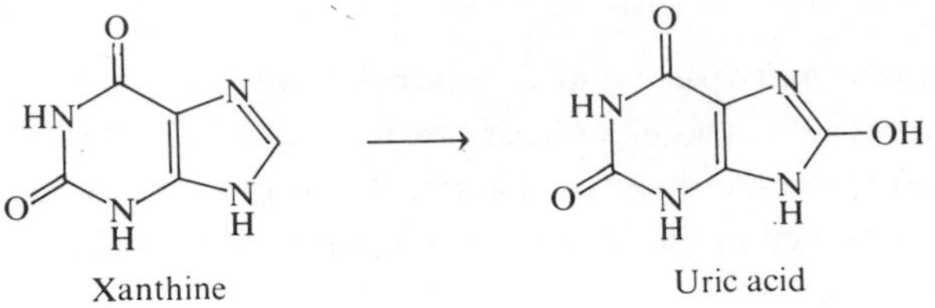

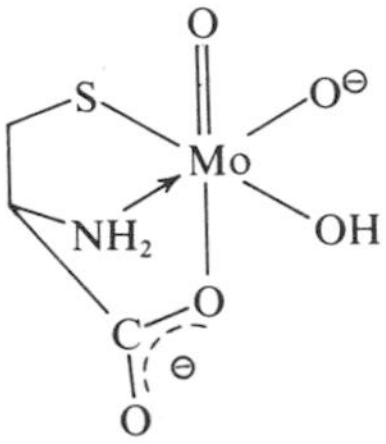

Xanthine   Uric acid

(6.20.1)

Figure 6.20. Xanthine oxidase: reaction and a model complex.

proposed with reasonable confidence that cysteine through its sulph-hydryl group is a probable ligand in the enzyme. Xanthine oxidase is an example of a wide range of enzymes that catalyse oxidation reactions which are very poorly understood because of complex inter-actions between their many components (ref. 58). These oxygenases generally require molecular oxygen as oxidant and always contain active metal ions (iron, copper or molybdenum) and very often elec-tron transferring cofactors such as flavins, pterins or ascorbate. A wide range of substrates including phenolic compounds, amino-acids, and steroids can be oxidised by the appropriate enzyme. The hydroxylation reaction exemplified in xanthine oxidase is of great importance today because it is one of the commonest pathways by which the body metabolises intruding foreign compounds such as drugs (ref. 59).

Model studies of such reactions are essentially sophisticated modifi-cations of the well-known oxidation reagent, Fenton's reagent (ref. 60) (figure 6.21) and use substrates known to be oxidised by enzymes. For example, iron coordinated by ethylenediaminetetra-acetic acid will oxidise quinoline, tyramine and many other aromatic compounds to a hydroxy derivative (e.g. 6.21.1) presumably via the highly reactive hydroxyl radical generated by interaction of iron(II) and hydrogen peroxide which itself is generated *in situ* from oxygen.

In the model, hydroxylation occurs at the most electron rich position but the enzymes control hydroxylation with respect to stereochemistry

$$Fe^{II} + H_2O_2 \longrightarrow Fe^{III} + HO^{\bullet} + HO^+$$

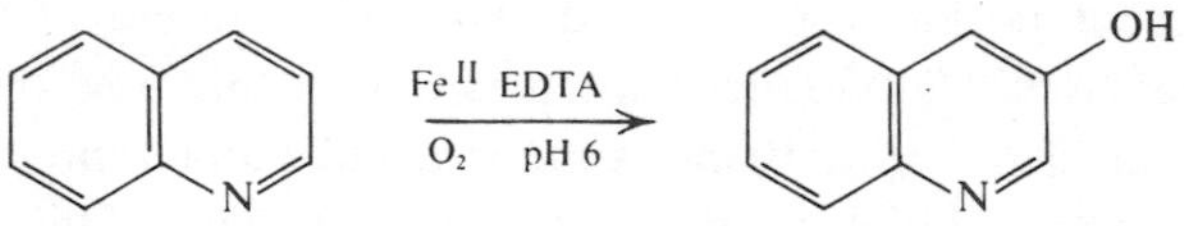

(6.21.1)

Figure 6.21. Fenton's reagent and model hydroxylation reactions.

as well as position. Recently progress towards mimicking this selectivity chemically has been made. The best mechanistic explanation of hydroxylase activity possible today is quantitative with respect to one functional unit of the enzyme at a time, for example the molybdenum(V) species and its reactions or the iron catalysed oxidation exemplified by the last model discussed. How the chain of redox processes which regenerate each active site after an oxidation or electron transfer operate is obscure.

In examining metalloenzymes by modelling comparisons, we have discovered more discrepancies between the behaviour of models and prototypes than in any group of natural reactions discussed so far. Of course, the variety of tasks to which metal ions are coerced for example as redox agents, conformation regulators or binding sites for small molecules, is greater than can be expected of a single amino-acid side-chain. Such a multiplicity of possibilities makes the drawing of generalisations more difficult. The problems in correlating models and prototypes in this field are symptomatic of an attempt to simulate a poorly characterised prototype and are aggravated by the subtle, concealed environmental effects which each enzyme appears to exert in a unique manner.

6.6 Simulations of the origin of life on earth

Continuing along our course leading from well-characterised enzyme prototypes for model systems through less well-charted enzymic territory, we ultimately reach a prototype which will 'perhaps' always remain unobservable, namely the conditions existing on this planet from which the molecules which comprise living cells emerged. The major source of information upon which to base model studies of *chemical evolution*, the name commonly given to the study of molecular evolution before the cellular or Darwinian evolution began, is geology. It can be estimated that about 10^9 years elapsed from the condensation of the earth some 4.5×10^9 years ago until the first cellular organism, the Gilbertian 'protoplasmic primordial atomic globule' and ancestor of us all, was evolved. This was the era of chemical evolution (ref. 61). All evidence indicates that at this time the earth's atmosphere consisted of methane, ammonia and water vapour and was chemically reducing rather than oxidising as it is today. This primitive atmosphere was under a constant bombardment by ultraviolet radiation from the sun, the protective belts of ozone in the upper atmosphere not having yet developed. Model systems attempt to simulate these conditions by subjecting dilute aqueous solutions of the postulated components of the primitive atmosphere, under a model reducing atmosphere, to intense radiation, of energy comparable to the solar energy of chemical evolutionary times. The energy source is usually electric discharge (ref. 61). From irradiation of such 'primeval soups' many molecules of biological significance have been isolated including many amino-acids (ref. 62), nucleic acid components (ref. 63) and porphyrins (ref. 64). The further development of the simulations to demonstrate combination of these biological monomers into their polymeric forms as peptides or nucleic acids has been difficult because laboratory experiments can scarcely model the vast time scale of the prototype system. However, the formation of nucleotides and peptides has been observed on irradiation of solutions containing the carbodiimide (6.22.1), which itself has been shown to be formed in model experiments under these conditions, and also on irradiating the normal 'soup' ingredients (ref. 65). The coupling process in these polymerisations is directly analogous to the well-known procedure of peptide synthesis in which dicyclohexyl carbodiimide is the dehydrating agent.

It is remarkable that some of the most primitive proteins which evolved soon after this period of chemical evolution in response to the newly developing oxidising atmosphere should have survived to the present day. From the common structural features of the cytochrome-cs of many species it has been estimated that this protein has remained essentially unchanged for 4×10^8 years (ref. 66). How these developments took place at the molecular level and the problem of the further organisation of biopolymers into cells are immense problems which can be approached only by modelling, a unique situation.

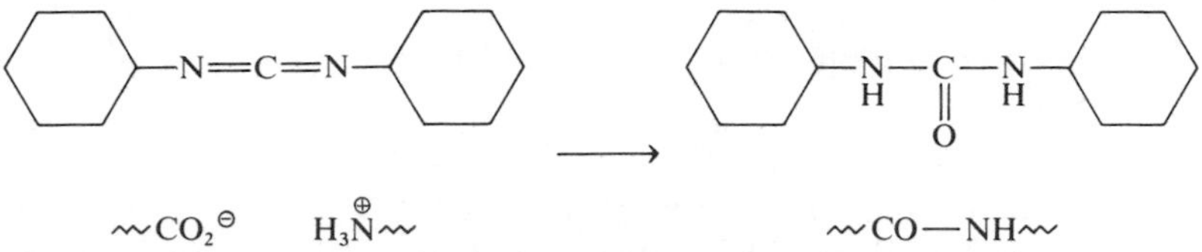

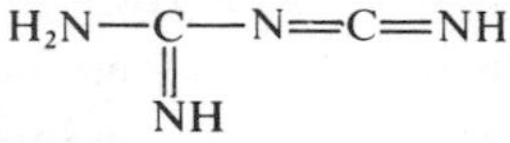

$$\sim\!CO_2{}^{\ominus} \qquad H_3\overset{\oplus}{N}\!\sim \qquad\qquad\qquad \sim\!CO\!-\!NH\!\sim$$

$$H_2N\!-\!\underset{\underset{NH}{\|}}{C}\!-\!N\!=\!C\!=\!NH$$

(6.22.1)

Figure 6.22. A possible chemical evolution of polypeptides.

6.7 Model studies of coenzymes

In striking contrast to the prototype of the preceding section, the *coenzymes* are the best characterised group of biologically important molecules involved in chemical reactions, except, of course, the substrates. This is largely because of the relatively low molecular weight of coenzymes (10^2–10^3) compared with enzymes (10^4–10^6 or more). Coenzymes bind to their enzymes usually by ionic interactions and are then able to react with the substrate directly under the influence of the enzyme. They are sometimes regarded as 'biological reagents'. Unlike a prosthetic group, coenzymes are not permanently bound to their enzymes but may dissociate and become involved with a different enzyme and a different reaction in another part of the cell. The precise chemical structures of most coenzymes are firmly established and this greatly aids mechanistic studies with the aid of models. When models of coenzymes are being discussed, there is a tendency to forget that an enzyme is also required for the reaction and this is to ignore an indispensable factor in their environment. Although, in practice, coenzyme models have proved successful in explaining the fundamental chemistry of reactions involving coenzymes without regard for the enzyme itself, problems can easily arise when this approach is applied to detailed mechanistic studies as we shall see. However, we can in general assign the enzyme to the role of controlling both position and stereochemistry of reaction.

6.7.1 *Nicotinamide coenzymes – model studies in physical organic chemistry*

The primary reducing agent of nature is a coenzyme, *nicotinamide adenine dinucleotide* (NADH) or its phosphate (NADPH) which supplies hydrogen for the reduction of many substrates but chiefly aldehydes and ketones (figure 6.23). The reverse reaction is also possible (ref. 4a). As can be seen from figure 6.23, the structure of NAD is rather complicated but it can be simplified for modelling purposes when it is recognised that the business end of the molecule is the *N*-alkylated nicotinamide. The first to take account of this was Westheimer (ref. 67), who chose a model system (coenzyme model and substrate model) such that the reaction rate was readily and accurately measurable under a variety of conditions. This was achieved with a *N*-benzyl pyridine derivative playing the part of coenzyme and a thioketone, more reactive than an ordinary ketone, as substrate.

With such well defined molecular modelling operations, the results

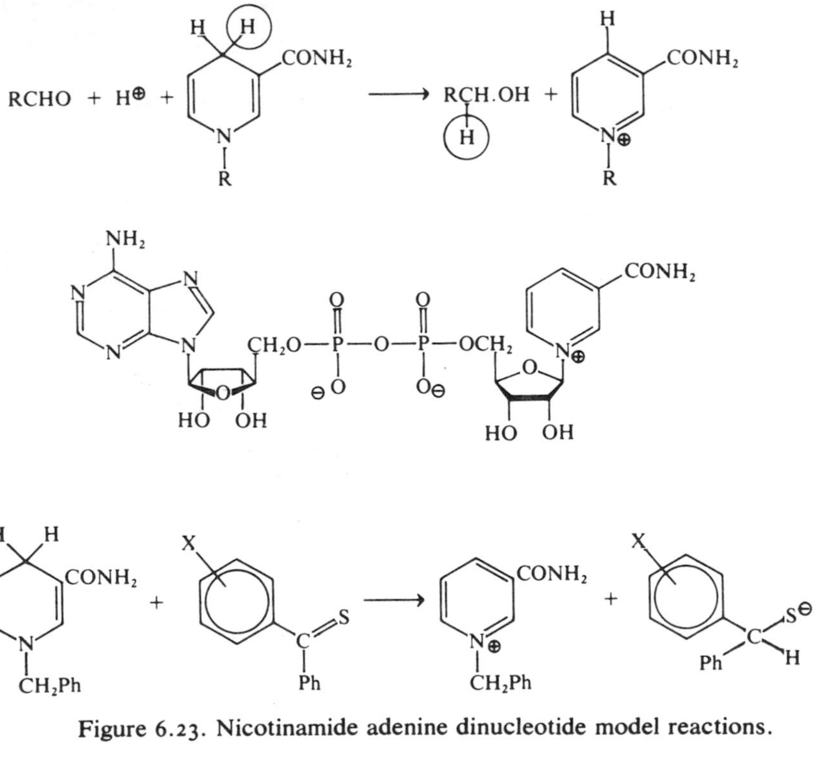

Figure 6.23. Nicotinamide adenine dinucleotide model reactions.

were easily related to the enzyme catalysed reaction. The mechanistic question to be considered is essentially: how does the hydrogen transfer take place between coenzyme and substrate? *A priori* it could occur in any one of three ways – hydrogen atom transfer, proton transfer or hydride transfer. Westheimer found a kinetic isotope effect in his reaction ($k_H/k_D = 4 - 5$) which is consistent with any of the three proposals occurring at the slowest step of the reaction. If a hydrogen atom were involved, then free radical trapping agents should inhibit the reaction but no such inhibition was found. The reaction was retarded by the presence of electron-donating substituents in the model substrate but accelerated by electron-withdrawing substituents. Therefore, it was concluded that a hydrogen transfer via the hydride ion accomplished reduction in the model reaction. Every subsequent result of model reactions and stereochemical studies of many enzyme catalysed reactions has led to similar conclusions (ref. 68).

Further simplification of the model system, to the greatest degree possible for a model–prototype correspondence to be defined, namely to 1-methylpyridinium iodide, produced results which correctly predicted a property that was subsequently found with the prototype in its enzymic context. It was observed (ref. 69) that the normal ultraviolet spectrum of the pyridinium cation was modified by the presence of iodide ion and the bathochromic shift observed (260 nm to 302 nm in water) was explained by a charge transfer interaction between the electron deficient aromatic ring and the polarisable donor iodide ion. Not only iodide ion but also other donors such as indole derivatives form charge transfer complexes with nicotinamides. In particular the spectrum of the latter are very similar to that of the NAD^+/glyceraldehyde-3-phosphate dehydrogenase complex from which it has been surmised that a similar charge transfer interaction plays some part in binding the coenzyme to the enzyme perhaps through a sulphydryl group which, like iodide ion, is also polarisable (ref. 70). Staunton gives a concise account of the salient features of NADH model chemistry.

6.7.2 *Pyridoxal phosphate – close correspondence*

In all cases in which the correspondence between the model and its environment and the natural system has been shown to be close in one respect, one may make cautious predictions from the model experiments as to the reactivity of the natural prototype, but predictions concerning detailed mechanism involve a considerable element of risk. Perhaps the widest correlations of enzyme and model reactions have been found with the reaction of pyridoxal phosphate (figure 6.24). This coenzyme binds to its enzyme through formation of a Schiff's base between the aldehyde function of the coenzyme and the ϵ-amino-group of a lysine residue in the enzyme; each enzyme adapts the further reaction of Schiff's bases to catalyse for example, transamination, decarboxylation, or transfer of a larger group (ref. 71). If the substrate is an amino-acid, the lysine Schiff's base is exchanged for a Schiff's base with the α-amino group of the substrate to form the intermediate (6.24.1) which is, of course, bound to the enzyme. The decomposition of the intermediate can then proceed in one of three ways – cleavage of bond (*a*) gives rise to decarboxylation, if bond (*b*) breaks, deamination occurs, and lastly a retroaldol reaction is the result of cleaving bond (*c*). All of these reactions and many others have been replicated with models, commonly using the unphosphorylated coenzyme itself (ref. 72). A straightforward correlation such as exists between this model and its prototype is very rarely attained.

6.7.3 *Coenzyme B$_{12}$ – a cautionary tale*

In contrast to the preceding example in which the model co-enzyme happens to be readily available from commercial sources, it would be astonishing if model studies are ever performed upon synthetic coenzyme B$_{12}$ or any of its derivatives (figure 6.25). The structure of this large molecular was inaccessible to chemical degradation and spectroscopy but was determined instead by X-ray crystallographic methods (ref. 73). Models have therefore been constructed from simpler ligands than the corrin (6.25.1) which is the fundamental ligand of the coenzyme B$_{12}$ structure. Reactions which require the intermediacy of coenzyme B$_{12}$ or closely related derivatives include methyl group transfer and molecular rearrangements (for example of diols, lysine and methyl malonate, see figure 6.25) (ref. 74). Branched carboxylic acids such as methyl malonate cannot be metabolised in mammals without the presence of B$_{12}$. If deficiency of the coenzyme exists, pernicious anaemia can result which is in part a consequence of the fact that methyl transfer or rearrangements are impossible without the coenzyme.

When the structure of coenzyme B$_{12}$ was elucidated it was found to contain a feature that had not previously been observed, namely a cobalt–carbon bond. The chemistry of this species was quite unknown

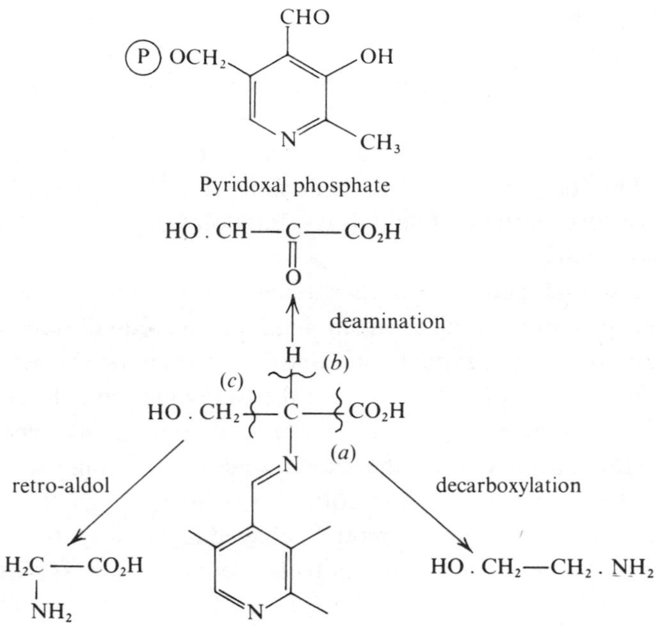

Figure 6.24. Some reactions catalysed by pyridoxal derivatives.

and this itself poses a problem for models to solve in conjunction with the usual problem of interactions with the enzyme and mechanism. The active structural unit in this coenzyme was recognised as the square planar chelated cobalt atom and a model must clearly incorporate this feature. Stable dimethyl glyoxime chelates of cobalt, cobaloximes (6.25.2) fulfil those requirements and they have been found to be capable of simulating many of the reactions which coenzyme B_{12} or its derivatives undergo in the presence and absence of enzyme (ref. 75). This is a consequence of the fact that, like the coenzyme, the glyoxime bound cobalt in the reduced cobalt(I) oxidation state is a very powerful nucleophile (ref. 76) capable of reacting with positively polarised carbon to form the catalytic intermediate cobalt–carbon bond. The metal–carbon bond may be compared with a Grignard reagent in which the polarisation is similar; both cobalt species, model and prototype, can show the typical reaction of methyl magnesium bromide and form methane by reaction with proton donors.

That this reactivity of cobalt in essentially square planar complexes has an important role in B_{12} reactions is beyond doubt. However it must be realised that the model has only one feature in common with its prototype, that is the cobalt atom chelated by four nitrogen atoms which are disposed in a square planar array around it, and by two variable ligands. Clearly, care should be exercised in extrapolating from a single correspondence model particularly on mechanistic points. The force of this caveat is demonstrated by a consideration of kinetic results on the diol-dehydrase reaction (6.25.3) in both enzymic and model cases.

The essential points of rather lengthy arguments were that the modeller proposed a mechanism implying an equilibrium between substrate and both enzyme bound and free coenzyme species (ref. 77) but the biochemist disputed this one the following grounds (ref. 78): if an equilibrium such as proposed from models were involved, then a tritium label introduced on the substrate should during the course of reaction become more or less equally spread between the substrate and coenzyme. Although the model behaved in this way, the prototype showed no tendency to divide its tritium equally, almost all the label being found in the products. The failure of the model mechanism to hold in the enzyme catalysed reaction serves to emphasise that a coenzyme is never free from the environmental control of an enzyme during reaction; neglect of this fact led to error. One error such as this does not detract from the success of cobaloximes in explaining the chemistry of the natural cobalt organometallic compounds. Every model has its limitations and it is up to the creator of the model to define the limits of relevance and work within them.

Models of other coenzymes, biotin, tetrahydrofolic acid and thiamine have been equally successful in explaining the reactions of their respective biological counterparts. The modelling procedures are identical to those described here and will not be further discussed. The reader is referred to Bruice and Benkovic's book for relevant examples, references and discussion (ref. 79).

6.8 Models relating to the nature of enzymic catalysis

6.8.1 Chemical properties of proteins

In our discussions of the applications of modelling to biochemical systems we have assumed a knowledge of the fundamental properties of all proteins including enzymes. The forces which govern protein structure in general can also be studied by models and *poly-α-amino-acids* have commonly been used for this purpose (ref. 80a). The conformational changes which can occur within proteins are well

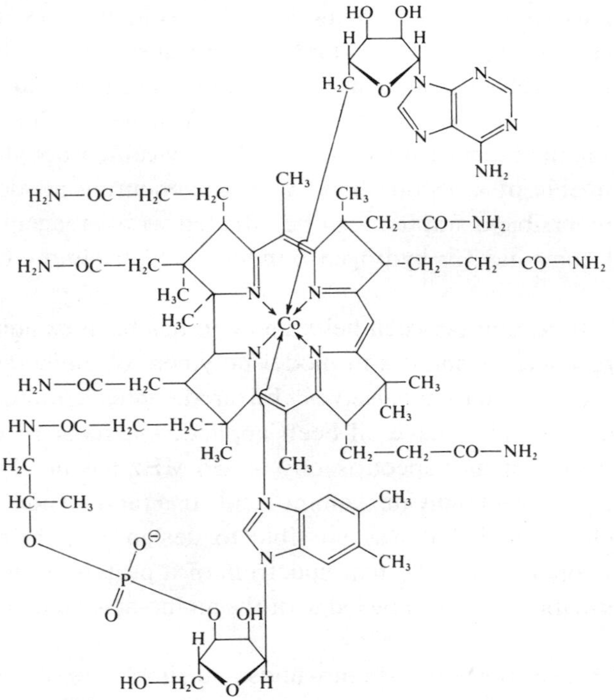

Coenzyme B$_{12}$

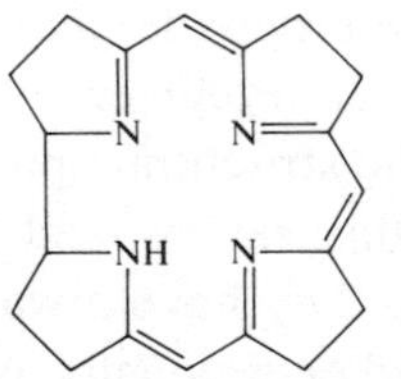

(6.25.1)

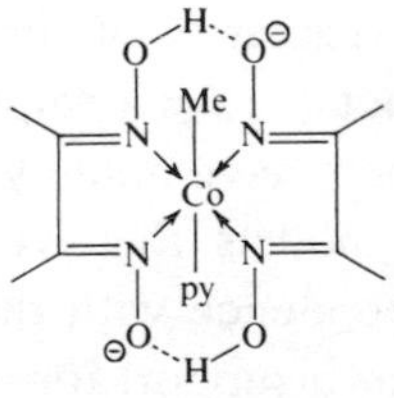

(6.25.2)

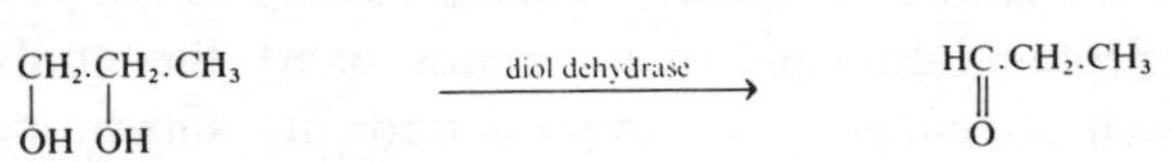

$$CH_2.CH_2.CH_3 \xrightarrow{\text{diol dehydrase}} HC.CH_2.CH_3$$

with OH, OH below the left structure and O below the HC on the right.

(6.25.3)

reaction with CH_3, $CO.SCoA$, H, C, CO_2H giving $CO.SCoA$, CO_2H.

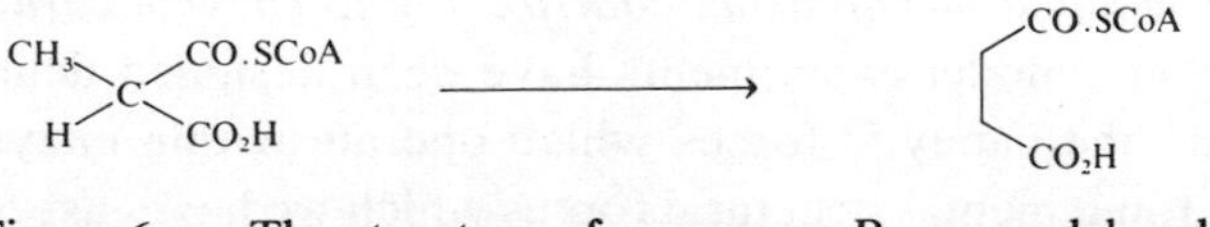

Figure 6.25. The structure of coenzyme B$_{12}$, a model and some reactions catalysed by coenzyme B$_{12}$.

suited to study with such models; for example, the reversible denaturation of a protein, in which its natural structure is distorted by a chemical or physical influence, occurs via a conformation change from the usual α-helix to a random coil conformation. This is a result of breaking the intimate hydrogen bonds between carbonyl and nitrogen components of a peptide bond with neighbouring peptide bonds and such reversible disruption can be effected with a reagent such as urea which competes for hydrogen bonding sites or simply by raising the pH.

This transition between helix and coil has been examined by many spectroscopic techniques in model polymers of amino-acids and also on natural proteins themselves. Infrared, optical rotatory dispersion and X-ray studies have all been applied, but most recently nuclear magnetic resonance spectroscopy at 220 MHz has been able to assign resonances from individual amino-acids in a random coil conformation in such detail that it was possible to design a mathematical model which could predict the nmr spectrum of a protein in the random coil conformation from a knowledge of the amino-acid sequence alone (ref. 81).

Other uses for poly-α-amino-acids and simple copolymers of amino-acids as models have been as substrates for peptidases (ref. 80*b*), and in studies of the genetic code (ref. 80*c*). It is clear that the wide applicability of such models is largely a consequence of the controlled regularity of the interactions within the peptide chain; this is unpredictable for complex natural polypeptides. Polymers of various monomers have notably been used as the basic structural unit of many of the models that we have examined in this review and in close correspondence with the prototype proteins, they have always functioned as a support for the reactive groups and occasionally as binding sites (refs. 34, 44). Whether essentially hydrocarbon (ref. 34) sugar or polypeptide in nature, polymers can be readily employed to model structural protein although they cannot exert the conformational strains which an enzyme can impose upon its substrate. We shall return to the subject of strain shortly.

6.8.2 *Chemical mechanisms contributing to enzyme catalysis*

Some model experiments have been designed deliberately to investigate the catalytic forces which operate within enzymes, aside from the fundamental structural forces which we have just considered. We have seen, in the previous sections, the chemical basis for many catalytic reactions by such mechanisms as nucleophilic catalysis (section 6.3) and approximation (section 6.3) for example, but alone this is insufficient to explain the extremely high rates of the enzyme catalysed reactions. Other effects must be operating. One such effect is a modification of the approximation mechanism we have already examined, that is that the reacting groups should not only be close together in space but should also be constrained by the enzyme to be in the optimum conformation for reaction. It is possible that the zinc atom in carboxypeptidase plays a dominant role in this connection as we have seen (section 6.5). Evidence that the magnitude of the effect of conformational direction can be exceedingly large has come from measurements of the rates of lactonisation of the substituted 2-hydroxyphenylbutanoic acids (6.26.1, 2) (ref. 82). The highly substituted compound with a constrained conformation (6.26.2) lactonises 10^{11} times faster than the unsubstituted model (6.26.1) at pH 7 with catalysis by imidazole. This freezing out of the optimum conformation for reaction has been discussed in detail in the case of the peptidase, thiosubtilisin (ref. 83) but controversy surrounds the precise interpretation of the factors involved (ref. 84).

Another effect which causes rate enhancements, hinted at a little earlier, is the induction of strain in the substrate and enzyme by their mutual interactions. This has often been cited as a driving force in enzyme catalysed reactions and has been demonstrated in a model system by Wang (ref. 85). It is known that many transition metal ions can catalyse the decomposition of the potentially bidentate hydroperoxide anion, HOO^-, and Wang chose to design a metal complex which would bind the anion in a bidentate fashion under considerable strain. This was achieved with the ferric complex of triethylenetetramine (6.26.3). The decomposition of the species with hydroperoxide ion bound to form oxygen was 10^4 times faster than the same reaction catalysed by methaemoglobin under the same conditions. Because in methaemoglobin the iron is bound by the square planar porphyrin chelate, hydroperoxide ion can only bind as a monodentate ligand and under no strain in consequence. However this large rate of reaction of the model system (6.26.3) is still 10^4 times slower than the rate of the natural reaction catalysed by the enzyme catalase, which may induce greater strains or, of course, use other catalytic mechanisms in combination.

Although in one enzyme catalysed reaction, a particular catalytic mechanism may be dominant, it is highly probable that the overall

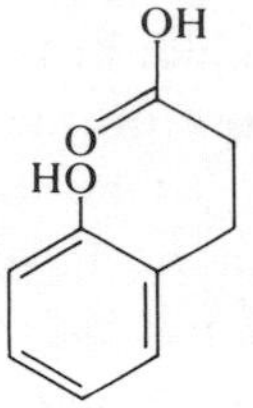

(6.26.1)

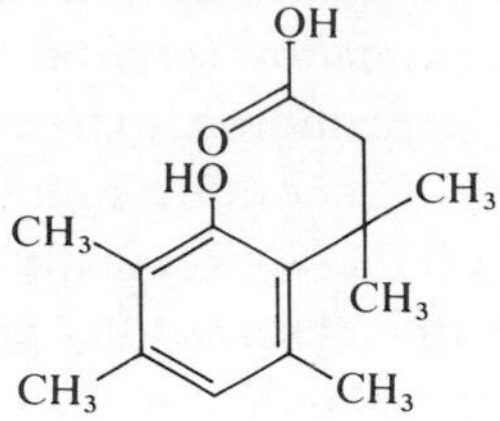

(6.26.2)

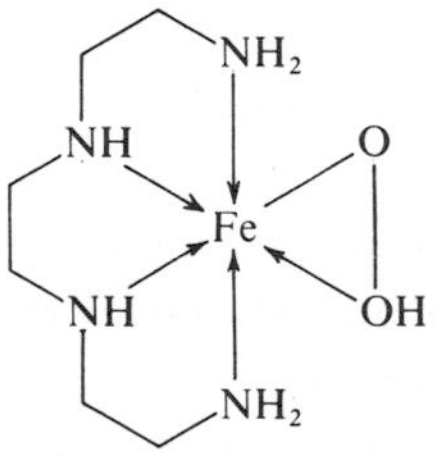

(6.26.3)

Figure 6.26. Model systems illustrating possible rate accelerating effects in enzyme catalysed reactions.

catalytic effect is a combination of many of the mechanisms we have discussed and illustrated by models. The enzyme mediates between the many possibilities through its most important property which studies of model systems emphasise, the control of the microenvironment of the catalytic site. It is noteworthy that the most successful model systems have been those which have included at least an approximation to the enzyme's internal environment.

6.8.3 *Control mechanisms*

In these chemical examples closely related to biology it is often necessary to express the results of experiments in mathematical form. Reference has already been made to one example of the use of mathematical models (ref. 80) but in reality, every kinetic expression of the results of a model run is itself a model, albeit a snapshot, of the reaction described in terms of well-defined kinetic parameters. The philosophy of this aspect of modelling has been discussed earlier in this book (chapter 4). If we choose to enlarge our frame of reference and consider the workings of enzymes in a wider metabolic and bio-synthetic context, it is obvious that the maintenance of balanced forms

of life, plant, animal and microorganism, depends upon the organisation and control of the many possible reactions within a cell. Two
of the basic mechanisms which can switch an enzyme catalysed reaction on or off are *repression of synthesis by the product of reaction
itself* (ref. 86) if it reaches too high a local concentration, or alternatively a compound further away in the metabolic chain may switch
off formation of a precursor by a *feedback mechanism*. The reverse
process may occur to stimulate production of a metabolite. Mathematical treatments of such systems are well developed (refs. 86, 87,
and 88) and the novelty here is the biological background. To bring
the concept of control to a molecular biological level, a mechanism
of control known as *allostery* deserves mention (ref. 89). The basic
hypothesis is that an enzyme can exist in two states or conformations,
one in which it is active, and one in which it is not. Such enzymes are
usually composed of two or more interacting subunits which can each
bind not only a molecule of substrate but also an activating or an
inhibiting molecule. The activator switches the enzyme on but the
inhibitor changes the conformation of the enzyme so that no further
reaction can occur. Commonly, the inhibitor or activator molecule is
fed back from a later stage in the metabolic sequence. For example,
the biosynthesis of L-threonine and L-isoleucine are connected by the
sequence shown in figure 6.27 (ref. 90). The first enzyme of the

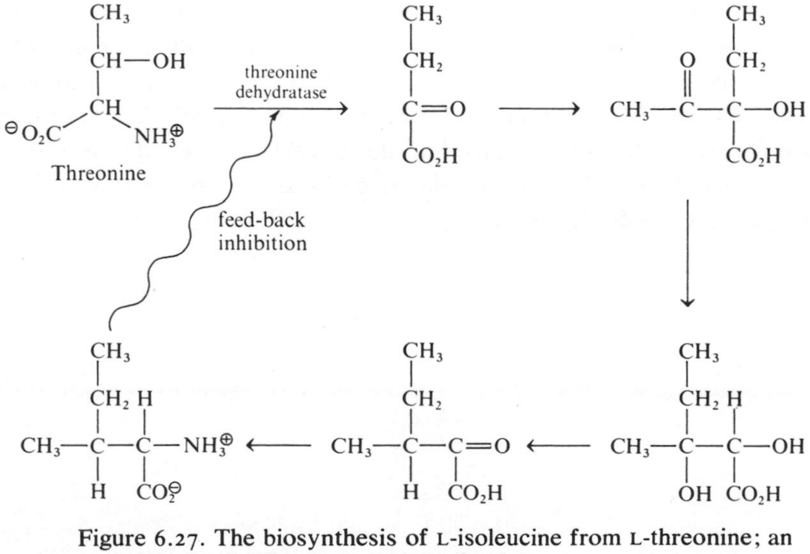

Figure 6.27. The biosynthesis of L-isoleucine from L-threonine; an
example of a feedback control mechanism.

sequence, threonine deaminase (E.C. 4.2.1.16), has been shown to require a threshhold concentration of threonine to be present before it will react, but on the other hand if too much isoleucine is formed, the isoleucine acts as an inhibitor of the deaminase and switches off the sequence. The importance of controls like this in maintaining the balance of amino-acids available for protein synthesis is obvious. Allosteric phenomena are associated with the interactions between various subunits of enzymes and proteins; this model of interaction has been found to be an adequate explanation of the oxygenation behaviour of haemoglobin which has two pairs of identical subunits (ref. 91).

6.9 Conclusion

In the preceding discussion we have seen the large part that model studies have played in developing understanding of a wealth of biochemical problems, but we have also noticed the limitations of model experiments particularly where environmental effects are operating. In the future, model systems must be refined to take greater account of environment correspondence with their prototypes if further progress in elucidating the mechanisms of enzyme catalysed reactions using this technique is to be made. On the other hand, so much having been learned about the factors which control enzyme reactivity, we are now in a position to reverse the direction of the modelling operation and, using the same approach as the biogenetic-type synthesis (section 6.2) to design our own catalysts for recalcitrant reactions based upon what we have learned about enzymes. Such catalysts might have industrial as well as academic significance. Bender's cycloamyloses are a first step in this direction. There has been much mimicry of enzyme catalysed reactions with considerable success but the capitalisation of the knowledge gained as synthetic catalysts or in other ways is a project for the future (refs. 92, 93).

7

DESIGNING CHEMICAL PLANT

7.1 The business as environment

It is through discovering new products and developing new manufacturing processes that the chemist makes one of his major contributions to society. But both product and process are potentially grave misfits in the environment. It is to the skill of the designer, as we argued in the opening chapters, that we look to ensure that artifacts fit well in their task environment, and so in this and the succeeding, final chapter we will discuss the design and the use of models first in process development and then in the introduction of new products.

To recognise that applied chemistry is, in these ways, a major factor for change and therefore a generator of pressure on the world around us is not to assume as a corollary, as many academics do, that pure research is an activity that can be conducted in isolation and one that does not impinge upon the outside world. We have learned that discoveries arising from curiosity oriented research may be the genesis of literally world-shattering inventions – Einstein's original thoughts on his equation $E = mc^2$ bear testimony to this. Moreover, the academic researcher, like the applied scientist, consumes resources and forms a part of a social grouping: his colleagues, his students, his family, his neighbourhood, his scientific societies, all of which will be influenced to some degree by his attitude to his work – to his research – and the time and effort that he puts into it. Checkland (ref. 1) suggests that action research (see section 2.7.2) in physics is unthinkable. It is only unthinkable if, by ignoring factors that have just been mentioned, the physicist considers himself and his research as a closed system. Analogies between academic and industrial situations and problems are often closer than one might expect.

7.1.1 *Input–output models of a chemical business*

A manufacturing process must obviously be at home in the context of the business of which it forms part. Some of the requirements that this paramount need imposes may be deduced from an input–output model of a chemical business. The model that is presented below, though adequate for our purpose, is simple and descriptive rather than quantified. Input–output models in a much more sophisticated form play an important role in economic studies, a good introduction to this type of work will be found in Shackle (ref. 2).

	A chemical business
Inputs	Outputs
Plant, equipment	
Raw materials	Products
Services	Services
Manpower	
Technology	Technology
Cash	Cash
	Environmental disturbance

The business consumes resources and these are the inputs, principally: plant and other equipment, raw materials, services in the form of energy and material and also time spent by public employees – police, postmen, etc.; a work input from employees and an input of technology – a mixture of science and know-how. These factors are not commensurate but they all cost money and the total cost is represented by the cash input.

The business earns money from the sale of the products and services that it produces and also, sometimes, by selling proven technology. It will inevitably create some kind of environmental disturbance, beneficial perhaps in providing employment and possibly even improving amenity, but sometimes damaging.

Money, of course, flows in the opposite direction to the goods and services that it buys. The cash that is an input in our model is that used by the business to pay for its operations; it flows out of the business's bank account. Like the other resources listed under inputs it is consumed in generating the outputs. It is obtained either by borrowing or from the depreciation and profit that will be components of the cash output that will be generated if the business is running well.

Overall cash flow is critical; if too much cash flows out then a business may become bankrupt even if it is making profit. An input–output analysis of cash flow might look like this (not all surplus cash would, of course, necessarily be held in a bank).

Bank balance

Inflow	Outflow
Loans	Interest on loans
Income from issue of shares	Dividend on shares
Grants from governments	Taxes
Income from sale of products, services, technology	Cost of buying resources to run and develop business
Depreciation	
Profit retained in business	

If cash outflow significantly exceeds inflow, as it may do if trading is bad, if over-ambitious expansion is attempted or in times of high inflation when much additional cash may be required simply to finance stocks of raw materials, of materials in process and of finished products, then the business may find it impossible to borrow enough money to carry on.

A cash flow model is one of a set of models that are subsumed in our model of a chemical business. If all the inputs and outputs in this larger model fail to operate smoothly or to adjust continually to a suitable balance, then the business cannot run. Smooth operation requires that the inputs be *available* where they are needed and *suitable* for their purpose. The outputs must be *acceptable*. With these requirements in mind we can begin to tease out the strands of interaction of process with environment and to envisage what will be demanded if misfits are to be avoided.

For example, applied to raw materials, availability and suitability imply the delivery to the manufacturing site, reliably and at an acceptable price of the required amount of each raw material of a quality suitable for the process. These requirements have to be planned, it cannot be assumed without investigation that they will be met.

Availability and suitability applied to cash bring to mind the need to obtain enough money when and where it is needed (cash is not necessarily freely movable between countries) at an acceptable rate of interest and with a convenient date for repayment – short term loans for example are not suitable for financing a project that may take ten years or more from start of construction to break-even. Similar analyses can be made of the other input factors.

Acceptability is an obvious prime criterion for outputs. Products must be acceptable to customers in the ways that the manufacturer, as customer, expects of his raw material supplies. Moreover, as we shall see in the final chapter, there are many gatekeepers other than the customer on the path to the market place whose co-operation is essential: regulatory authorities for drugs are a case in point, and professional experts can be decisive for or against acceptability, for example in the case of chemicals that are used in fire-fighting or building. The plant and process will have to be acceptable to local authorities, to the factory inspectorate, to one's own employees.

Environmental disturbance is included among the outputs because the acceptability of the new process, both in regard to the working environment that it itself constitutes and its effect on its outer environment, are nowadays recognised as critical factors in the decision as to whether or not a process *is* acceptable.

Thus, through an input–output model of a rather hybrid type, we can explore many of the factors that favour or hinder the introduction of a new process. The proposition to build the new plant will imply a commitment over a long period, 20 years or more, from design, through building, to operating and maintaining the plant and appropriate resources must be available for every stage.

7.1.2 *The faces of time*

The long time scale of developments in the chemical industry creates, as it does in many other industries, especial problems for the planner. The problem is not simply to build dynamic models but to make the necessary predictions, social, economic and technological, as to what changes will occur over the lifetime of the project.

Not only do the values to be ascribed to variables change – that this should happen is in the nature of variables – but the list of relevant variables will change, some losing their significance and other factors unexpectedly intruding.

Time exerts itself in chemistry primarily as the denominator in kinetic equations; as the dimension in which rates of reaction are measured. In applied science time is also the inelastic dimension in which to do things. Have we enough time? Since we have all the time there is, the question must be – can our project move fast enough relative to other concurrent events and processes – money running out, competitive initiatives and so on?

But sequence as well as speed is important in the time dimension. Are all elements of the development well integrated in time? Will the various steps in design, procurement, building, start-up, etc. be so phased as to ensure a smooth run through? Modelling techniques such as bar charts (ref. 3) and critical path analysis (ref. 4) can be very helpful in this connection.

Time is needed also to adapt: for the project to adapt to its environment and for the environment to adapt to it. Too great a rate of change can produce disabling stress in people and in their physical environment – the comments in chapter 2 on Simon's concept of artificiality have taken this point further. Many changes that would be acceptable and even welcome if made gradually, arouse strong opposition when pushed too fast.

Time, or the wear and tear that it imposes, is also, for the applied scientist, something to be withstood – which means in this context that the rate of those physical and chemical processes, such as corrosion of structures and equipment and loss of activity of catalysts, must be acceptably slow.

And time, for the scientist, is often for being patient in and for everyone, for living in.

The applied scientist, like the economist, is concerned fundamentally with dynamic systems. He must study not only states of affairs but how to reach them, remain in them and move, when necessary, out of them.

7.2 Problems of process design

In the choice of a new process to be developed to full scale, several alternatives will usually have been tried in the laboratory and tested by the model that was described in the previous section, or by some equivalent procedure.

The problems will be the less if the task in hand is to design a plant based on an existing and tried process than if the plant incorporates significantly novel technology. If the new plant is to be, as is often the case, bigger than has previously been built, then scale-up problems of the kind that will be described in the next section have to be faced, as to a greater or lesser extent have those problems posed by a new environment.

An especial danger will be the temptation to make minor changes calculated to improve performance without thoroughly thinking through the implications for the plant as a whole. A chemical plant is a system very rich in interactions and ill-considered alterations can have unexpected and potentially disastrous results.

That costs fall as plant size increases and experience on operating a process accumulates has been well established and the Boston Consulting Group have quantified the progression and expressed it as an experience curve (ref. 5). If a new process is chosen, then a start is made on a new experience or learning curve, with considerable opportunity for improvement – and a break with the conventional wisdom which, if a calculated risk and not an impetuous jump into the dark, if informed judgment rather than best guess, can pay off handsomely. Nevertheless the financial case for building a new plant to replace and shut down an old one is sometimes particularly difficult to make because a well-maintained plant that has been operating for ten to twenty years should have progressed far along its learning curve and be running near its best. Moreover its capital cost will probably already have been written off, that is to say covered by the sums set aside annually as depreciation so that the absence of a continuing charge for depreciation will significantly reduce product cost. To justify spending money to replace the old, the new plant will have to show savings and these typically come from improved efficiency in resource utilisation, that is to say through economies of scale or through more efficient processes. Although debottlenecking exercises are frequently very profitable they must not be taken too far. Learning curves going back many years show a reduction in rate of learning during prolonged periods of plant extension in which there was no replacement of old plant by new.

At the stage in design that we are now considering, a substantial amount of chemical and chemical engineering information will have been assembled and the remaining task will be to design a safe, acceptable and effective manufacturing plant. Broadly classified, the potentialities for misfit with its environments (physical, social, business) are as follows:

(1) The plant does not work properly – it does not start up smoothly and make the desired product in the desired quantity and quality continuously and without undue maintenance.

(2) It is hazardous – it causes injury through e.g. poisoning, burning, crushing. The assault may be acute following an accident or chronic as the result of an ongoing unsatisfactory state. There is undue stress on operators.

(3) The plant pollutes. The boundaries of the hazardous and the polluting obviously overlap. Pollutants can conveniently be defined as disagreeable impacts on the environment such as noise, emissions

into the atmosphere or into rivers, an increase in traffic and so on which, as with hazards, may be chronic – a concomitant even of normal plant operation – or acute, following an unplanned event, that is, an accident.

(4) The process is wasteful, that is to say inefficient in resource utilisation, a situation that may arise through the wrong choice of process or of process conditions or through inability to control them.

(5) The process is uneconomic. The question of whether a process is or is not economic depends on the point of view. What is economic for the individual may be uneconomic for the business, what is economic for the business may be uneconomic in national terms, and what is economic for one nation may be wasteful or in some people's eyes unjust for the world as a whole.

These broad classes of misfit can be broken down into sub-classes and then into specific problems of smaller or greater potential danger, just as the plant can be decomposed, for the purpose of design and of construction, into quasi-independent units. In the whole process of searching out misfits, designing them out, of decomposing and re-synthesising the system into a working plant, models are used extensively not only for the idea generation and testing, and the fact finding that are essential in the designing of a complex system, but for communication at all stages, for training, for planning construction, for operating and for improving performance.

7.2.1 *Process development*

Models of many kinds are used extensively in process development and it is not possible to treat them exhaustively within the scope of this book. Our task is, in the context of the general principles of modelling that have been developed in earlier chapters, to sketch the types of problem that have to be dealt with and to introduce the various types of model. In accordance with the objective of providing practical assistance wherever possible, the final section of this chapter reviews the progress of the use of mathematical models and the computer in process development and provides references to texts in which extensive and informative expositions can be found. A useful and fairly comprehensive introduction to process development and scale-up will be found in Baines *et al.* (ref. 6), see also Ball *et al.* (ref. 7).

The task of process development is, through experimentation in the laboratory and probably on a semi-technical (pilot) plant, to provide enough technical information to ensure that the full scale plant is not a misfit but fulfils its purpose. The transformation that is required has two main components: (1) scale-up and (2) transfer to a different environment.

7.2.2 *Scale-up*

In some respects, operating on the larger scale may be less demanding than running a laboratory model, for example the accurate control and metering of flow-rates is usually simpler, but in many ways the task is more difficult and the first need is to identify those factors that are likely to become more demanding on scale-up.

The first obvious factor is that on a full scale plant larger quantities of chemicals will be handled than in the laboratory, so that an untoward occurrence – an escape of chemicals or an explosion for example – may have consequences that cannot be easily contained. In the laboratory, workers can be protected fairly easily but on the plant it may be quite another matter. Techniques for investigating potential hazards and operability problems have been developed over the last few years (ref. 8). These are introduced at an early stage of development with the aid of the various models that are used in the plant design work e.g. flow sheets and line diagrams. These procedures make an important contribution to safety and to sound operation. No-one concerned with designing plant should be unaware of them.

The need to handle larger quantities of material brings in its train the question of materials of construction. In the laboratory, glass is often a very suitable material for apparatus, being resistant to most chemicals and yet a sufficiently good conductor of heat and strong enough for use in small pieces of equipment. Glass can, of course, be used on the larger scale, either by itself or as a vitreous coating for metal vessels or pipes but in this situation heat transfer and thermal expansion can pose problems and there is always the difficulty of repairing the vessel if it becomes slightly chipped, which can often not be done *in situ*. The containment of pressure is another matter that may present a different face on the full scale plant.

The question of sizing can itself be a matter of some difficulty. It is obvious that dimensions cannot simply be scaled up linearly. If we increase the radius of a cylindrical reaction vessel three-fold, then the volume will be increased by a factor of nine but the side wall area only by a factor of three. It might well be that the reactor walls are jacketed with cooling water and used to remove heat from an exothermic reaction that is taking place in the body of the reactor. In this case, the surface available for heat transfer per unit of heat evolved would be, in the larger vessel, one-third of that in the smaller.

Failure to recognise that dimensional similarity did not apply in all respects, led to a serious fire (but no injury through good fortune, and because of proper precautions and fire drills) on a plant on which one of the authors was once working. An exothermic reaction (the dimerisation of pyridine by metal powder) was being carried out in a cylindrical vessel that had been scaled up from the semi-technical plant by simply increasing the depth so that the volume and side-wall area were increased proportionally which was convenient since much of the heat of reaction had to be removed through a water-cooled jacket encircling the side walls. However, some of the heat of reaction was to be removed by vaporisation of pyridine, which then condensed and ran back into the reactor. What had not been taken into account was that the pyridine vapour had to pass across the liquid–vapour interface and this, of course, was not scaled up when the reactor depth was increased. In the scaled up plant as the reaction proceeded the mixture became viscous and as pyridine vapour, in quantity proportional to the heat generated in the increased depth of reactor, tried to tear through the surface, large globs of sticky reaction mixture were carried into the condenser where they solidified. This blocked the outlet and pressure in the reactor built up and blew a gasket. The escaping spontaneously inflammable mixture ignited. This was an instance in which an important parameter, the rate of gas flow through the liquid–gas interface, had not been recognised. Fortunately it was on a model – in this case the semi-technical plant – that the lesson was learned.

The problem of simple dimensional scale-up has other more recondite aspects. For example, in catalytic processes the rate of diffusion in and out of pores may be a limiting factor, so that scale up will not simply be a matter of increasing the size of a reactor. In designing a column for distillation, one may forget that as the height is increased so the dead weight on the packing increases and this can lead, and has led, to crushing of the bottom layers of the packing. This problem, once recognised, can be easily dealt with by inserting intermediate supports.

Whenever possible, chemical engineers use dimensionless numbers to help them in scaling up. It is known, for example, that heat transfer between a liquid flowing in a pipe and a coolant or a heating material outside, is much easier to calculate if the liquid is in turbulent flow rather than flowing streamlined along the pipe. The criterion for turbulent flow is that the Reynolds number (the product of velocity and pressure head divided by fluid viscosity) should exceed a value of 400 to 800. As inspection shows, Reynolds number is dimensionless.

Reynolds number $= Du\rho/\mu$

where

$D =$ inside diameter of pipe,		(l)
$u =$ average velocity of fluid flow,		(lt^{-1})
$\rho =$ density,		(ml^{-3})
$\mu =$ viscosity.		$(ml^{-1}t^{-1})$

The expressions in brackets are the dimensions, where l = length, m = mass and t = time.

A helpful insight into the use of dimensionless numbers can be found in a paper from the Delft Hydraulics Laboratory (ref. 9) on the use of models in studying rivers and harbours. This paper and another from the same laboratory (ref. 10) contain in addition useful comparisons of iconic and mathematical models of the same system. Of the many interesting themes in these papers, two are particularly relevant to the practice of modelling in chemical engineering.

The transport and deposition of sand is obviously of fundamental importance in studying sand banks, silting and the like. How should the size of the particles of sand be reduced for the model? It turns out that grain size must be reduced on the same scale as the water depth if similarity conditions are to be met. This reduction can safely be applied to the coarser grains, but application of the same factor to fine grains would bring the size of some particles in the model to below 200 μ at which point the transport behaviour of sand particles in water changes sharply.

This difficulty can be circumvented by using a material such as bakelite which has a much lower relative density than sand, so that weight, the essential parameter, can be scaled down while keeping particle size above the level at which anomalous behaviour sets in.

This example illustrates how properties other than size may have to be altered in order to achieve similarity – in this case in scale *down*. Similar artifices are needed in chemistry for example in fluidised beds and in mixing fluids by jets when, on a small scale, viscosity may need to be changed in order to model the mixing patterns to be expected on a large scale.

A second example from Delft (ref. 10) illustrates the recognition of a factor that might otherwise be overlooked, the importance of assessing whether ignoring a factor is likely to lead to under-design, and the need to evaluate artifices used in modelling from all points of view.

Bubbles of air always form in a rough sea and cushion the eroding effects of waves beating against the land. These bubbles are not properly reproduced in models though Schoemaker (ref. 10) suggests as an experiment (apparently not yet tried) the operation of a model under reduced air pressure. Fortunately the lack of bubbles under model conditions aggravates erosion rather than reducing it so that, if the object of modelling is to design protection against erosion, a safe answer is likely to result even if the bubbles are ignored. Although the model may falsely suggest that over elaborate, unnecessarily expensive protection measures are necessary, inadequate protection with perhaps the risk of flooding is unlikely to result. The second factor also illustrates that there are problems in scaling down as well as in scaling up. Surface tension, though of negligible importance in the behaviour of waterways, becomes increasingly important as scale becomes smaller. In too small models, damping of waves by surface tension may be sufficient to make the experiment irrelevant to prototype conditions. The reader may have thought of adding detergents in order to reduce surface tension. Schoemaker draws attention to an interesting pitfall that may arise when this is done, pointing out that the reduction in surface tension is due to the concentrating of molecules of detergent on the surface in a special orientation caused by the electric polarity of the molecules. But, when in waves the surface is periodically extended and contracted molecules get in opposite orientation with the opposite effect. In the thin fluid layers around soap bubbles the same happens. Other disciplines' models are often a fertile source of ideas and analogies.

7.2.3 *Change of environment*

The second class of problem that is met with in transfer from laboratory to factory, namely the movement to a new environment, involves human and material factors; there will be changes in culture, skills, priorities as well as in materials. For example, those who operate the plant in the factory will be skilled at that task but they will not necessarily be good at experimenting, indeed it is unlikely that they will be so and therefore it will not be possible to rely on the operators to take care of variations in behaviour of the kind that might be easy to cope with in laboratories. Moreover, as we have already said, the consequences of a mishap in the factory may be very much more serious than they would be in a laboratory. The factory environment will provide inputs into the plant in the form of raw materials which may have been made on some other process in that factory, or usually services such as steam, water and the like, and a supervision from the factory management. All these demands will need to be worked out and be regarded as part of the operating model.

7.3 Process diagrams

The first formal description of a process is likely to be a line diagram, that is a schematic representation of the major units with lines connecting them to indicate the flow of materials. Figure 7.1 is a simplified line diagram for a process to make methylene chloride by chlorinating methyl chloride. The product, a mixture of unchanged methyl chloride, methylene chloride, chloroform, carbon tetrachloride and larger molecules formed by condensation, will pass to three distillation columns, the first of which will separate unchanged methyl chloride from the more highly chlorinated components, the second will strip off methylene chloride, and the third, chloroform. Each box represents a group of components. The still, for example, will comprise at least boiler, distillation column, condenser, reflux divider and requisite pumps, pipework and valves and instruments.

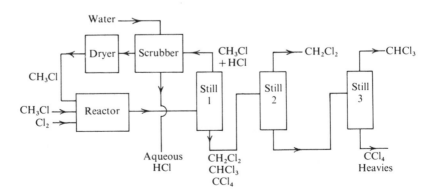

Figure 7.1

The unconverted methyl chloride returning to the reactor has to be freed from hydrogen chloride by scrubbing with water, to give usable aqueous hydrochloric acid, and then dried. The process, though designed to make methylene chloride, produces three other usable streams, aqueous hydrochloric acid, chloroform and a mixture of carbon tetrachloride and 'heavies', i.e. high-boiling residues such as hexachlorobutadiene, hexachlorobenzene and hexachloroethane. The heavies, incidentally, could probably best be used by including them in the feedstock of a plant designed to totally chlorinate at high temperature to a mixture of carbon tetrachloride and perchloroethylene, $CCl_2 = CCl_2$; a useful plant for processing all manner of chlorinated by-products that would otherwise present a disposal problem.

This is a relatively simple plant, though the design of the reactor may be quite complicated; rapid mixing of the gases is likely to be important as is temperature control of the highly exothermic chlorination. Provision for heating up the reactor in order to start the reaction will be necessary – a reminder that the plant designer can rarely limit his thinking to a steady state, normal operation but must deal with the dynamic problems of reaching it and leaving it – start-up and shut-down – as well as with variations during running. The quantitative flows of materials entering and leaving each vessel could be shown on the line diagram but this can lead to a cluttered presentation and so a flow sheet is usually drawn up on which, in parallel vertical columns the inputs and outputs from each unit are specified. All materials and energy flows can be listed: reactants, products, hot and cold water, nitrogen, etc.

This procedure relies on the fact that a chemical plant can usually be considered as a set of quasi-independent units with the output from one becoming the input of the next. There may be material feed-back, such as the return of unconverted material from a separating stage to a reactor and this can easily be incorporated in the diagrams. It is, therefore, very easy to decompose the model for design purposes but the real and fundamental interdependence of the whole system must never be lost sight of.

The objective at this stage is to design a plant that will achieve an acceptable performance at the least cost, that is to say at the lowest sum of capital and operating cost. In the equations that are set up to describe plant performance, cost will be the variable to be minimised. The standard of performance – product quality, output, etc. and, of course, safety will be built in as constraints. There will be many ways, often conflicting, of minimising the consumption of resources per unit of output. For example, increased expenditure on the reactor could ensure a better yield on raw materials consumed. Similarly the required purity might be achieved from an expensive reactor which gives a purer product and thus demands a simple separation unit, or one might go for a simple reactor and a complex and therefore more expensive separation. There will therefore be multiple interactions and many 'trade-offs' to be investigated, and this type of optimisation is particularly suitable for mathematical modelling and computer processing.

In the space available we cannot deal with models incorporating cost functions but in section 7.4 we give, by way of introduction, a simple example of how a mathematical model of a process is begun.

7.3.1 *Providing the data*

What has to be done now is to put flesh on the skeleton of the line diagram and to specify, with increasing precision and for each unit of the process, first a quantified duty, that is to say the reaction, separation or heat exchange that is to take place, and then a design that will perform that duty at minimum cost, defining *inter alia* size, shape, materials of construction and lay out.

The requisite data may be available from the work already done in developing the process or it may be found in the literature. The chemical engineer uses not only the usual scientific publications but also equipment manufacturers' brochures and catalogues. The suppliers of special items of equipment such as pumps, centrifuges, heat exchangers, furnaces, column packing and the like have a fund of knowledge on the use of their products and will frequently run experiments to find the best solution to a customer's requirements.

Care must be taken to ensure that data taken from the literature is relevant to the purposes in hand. Published reaction rates, for example, usually relate to conditions that are very different from those that obtain in a large reactor and can often serve only as an initial guideline (ref. 11). As we shall see in section 7.5, the problem of provision of data is now being greatly simplified by the assembling of data banks stored in computers.

It is, however, of the nature of innovation to require new data and this will usually lead to experimentation in the laboratory or on a semi-technical plant. Such plants are expensive to build and to run and the need for them is often debated. Majority opinion is probably strongly in favour of including a semi-technical plant among the design models because the consequences of failure at full-scale may be very serious. The semi-technical plant tests the projected extrapolation from laboratory to full scale both quantitatively and qualitatively. Qualitatively in that it provides another opportunity for things to go wrong, that is to say for inadequacies in our model to show themselves, while it is still relatively easy to put them right.

Once the flow-sheet, with its mass and energy balances has been firmed up, the line diagram can be finalised and individual units defined in detail. Cost models will have been used to assist in choosing between alternative variants, at first perhaps with unit processes costed approximately from their size and duty, the so-called factorial costing, and later on the basis of suppliers' quotations. In addition to the usual drawings a physical model may be made to assist in visualising the layout. As we shall see in section 7.5, all these procedures can be incorporated into mathematical models to which activity the paper *Digital computers in the design of complete chemical processes* (ref. 12) provides a very useful introduction with practical examples.

We close this section with mention of two situations which exemplified the imperative need to seek out all relevant factors.

A suitable material of construction was being sought for a reaction involving aggressive components, and the corrosion rates were measured in the usual way by partially immersing strips of metals in the boiling components. This procedure takes care of the fact that corrosion is often most severe at the air–liquid interfaces. Fortunately, it occurred to the experimenter to ask whether the rate of corrosion was likely to be affected by thermal stress, since the reactor was to be cooled by an externally fitted jacket, so experiments were done in which surfaces were cooled on one side and exposed to the reactants on the other. It turned out that some alloys which had been thought to be satisfactory from the standard tests were significantly corroded when the outer side of the test specimen was cooled. The accepted test procedure having proved to be too simple, a new standard that included more parameters had to be substituted.

Envisaging the plant as a series of connected unit processes is an important step in establishing quasi-independence. The outputs from one unit are the inputs to the next. Once again the qualifier 'quasi' must be emphasised. There are many examples of unexpected interactions between one unit and another; a particularly trying time was had by those running a polymerisation plant which was using specially purified water from a copper still. It turned out that minute traces of copper inhibited polymerisation. This might not have been too difficult to sort out were it not for the fact that the effect only occurred within a narrow range of oxygen content. The polymerisation was, in fact, intended to be conducted in the absence of oxygen under a blanket of nitrogen but fortuitous ingress of air through the gasket sealing the reactant vessel occurred and, of course, the rate at which it happened tended to change every time the reactor was opened up and resealed.

7.4   Mathematical models

For the chemical engineer the availability of computers has revolutionised the procedures used in designing chemical plant. The mathematical modelling, though fundamentally simple, is extremely subtle and complex in practice and it is not uncommon to have several hundred equations that cover not only such physical and chemical parameters as rates of reaction and of matter and heat flow, but also cost. To give any substantial introduction in the space available here is not possible. The interested reader will find useful presentations in Franks (ref. 13) Savas (ref. 14) and Rose (ref. 15) and in other texts that are mentioned in section 7.5.

A worked example that illustrates some of the ways in which the costs of alternative processes can be identified and compared is given in Baines *et al.* (ref. 6a) where the potential profitability of a piece of research into improving a catalyst support for an existing gas phase process is examined.

For those, on the other hand, whose need at present is a very brief and simple indication of the procedure, the following theoretical dynamic model for a first order, batch reaction $X \to Y$ may serve.

A reaction in which there is one, first order irreversible step

$$X \xrightarrow{\ k\ } Y,$$

can be described mathematically by the differential equation

$$\frac{dy}{dt} = -\frac{dx}{dt} = kx \qquad\qquad 7.1$$

where x and y are the molar concentrations of X and Y respectively at time t. Integration, setting the concentration of X at time $t = 0$ at x_0, gives

$$\ln x = \ln(x_0 - y) = \ln x_0 - kt. \qquad\qquad 7.2$$

This model, which plots as the straight line relationship of first-order kinetics, relates the concentration of X to the rate of conversion of X to Y at a given fixed temperature. To take account of the fact that reaction is dependent on temperature we can use the Arrhenius equation

$$k = Ae^{-E/RT} \qquad 7.3$$

which expresses the temperature dependence of k in terms of the frequency factor A and the activation energy E which are constants for a given reaction. Substituting for k in the previous equation we obtain

$$\ln x = \ln (x_0 - y) = \ln x_0 - Ate^{-E/RT} \qquad 7.4$$

which is a model that describes quantitatively the effect of the concentration of X, the reaction time, and the temperature on the rate of formation of Y.

The two factors that, in addition to rate of reaction, are of greatest general importance in chemical processes are heat and mass flow and these must be represented in the process model.

If the heat of reaction in the conversion of X to Y is ΔH joules mol^{-1} then, using the H negative convention for exothermic reactions, the rate of heat production from a volume V of reactants is $-V\Delta Hkx$. The rate of heat removal in a jacketed reactor is $vW(T-T_B)$ where v is the heat transfer coefficient, W is the area of reactor wall that is available for cooling and T_B the temperature of the cooling bath. If the specific heats of X and Y (at constant volume or pressure as the case may be) are C_x and C_y (J/mol °C) respectively, then the rate of temperature increase of the reaction mixture is given by:

$$\frac{dT}{dt} = \frac{-V\Delta Hkx - vW(T-T_B)}{c_x V_x + C_y V_y}, \qquad 7.5$$

which, together with equations 7.1 and 7.3 constitute a dynamic kinetic model for the reaction X → Y which would commonly be used in the analysis of batch reaction systems.

The equations that have been used, as is almost always the case with models, contain simplifying assumptions, for example that the activation energy is independent of temperature. As always it is necessary to make sure that the factors that are eliminated by these assumptions are not relevant to the situation in hand. If the assumptions are not justified, further relationships may be added and the resulting equations solved simultaneously with the former set. The limitation at this stage tends to be availability of data rather than the number of equations to be solved. It is sometimes convenient to take account of a variable by averaging; for example, a significantly better model might be obtained by taking the temperature of the water jacket as the average between that of the inflow and that of the outflow.

Mathematical process models can be used in designing plant, in operating it and in optimising its performance. In the design stage, models may include among the variables such parameters as reactor volumes and distillation column heights that will become invariant as the design is firmed up. Where general purpose plant simulation packages are employed the models usually operate in performance mode – that is a plant design is specified, feeds and operating parameters are set, and outputs from the plant, including costs, are calculated. Design of plants using such models is an interactive exercise between the designer and the model. General purpose, equation-based (as opposed to modular) systems do exist and these do not distinguish between design and performance parameters, but such systems are rarely used outside universities. Other parameters such as composition of the reacting mixture, or the nature of the column packing will remain variable to a greater or lesser extent. The establishing of the extent to which operating variables can safely be allowed to vary is a major task in process design, as is that of ensuring that the process will be sufficiently robust to stand changes in those variables which are uncontrollable, for example, wind strength on cooling towers, or deliberately uncontrolled, for example, ambient temperature.

A mathematical process model can be the basis of control by computer. This can either be on-line or off-line. In the former case the computer, having received and processed information from the instruments that monitor the state of the plant, itself activates valves and other control equipment in such a way as to maintain the plant within the desired operating conditions, in the same way as an aeroplane is flown by an automatic pilot. The model built into the computer to enable it to predict behaviour and to make the necessary changes may be theoretical or empirical or partly empirical. The advantages and disadvantages of these bases were discussed in chapter 2. Though a plant may be started up with what is essentially a theoretical model, change over to a well-based empirical model may be both economical in demand on the computer and very effective in operation.

Though aircraft pilots complain of boredom when flying long sectors on their automatic pilot, the introduction of computers to chemical plants has generally been seen by process operators as an extension of interest and challenge. Often the process operator learns from the computer in that he observes the modifications to the operating conditions that the computer makes and builds them in to his own personal model. Conversely the plant operators and designers can, as experience grows, often rethink the model in the computer and so there is a beneficial interaction.

It is of course possible to design a statistically based programme for optimising plant performance and this can either be incorporated in the on-line computer programme or operated off-line. A word on the dangers of optimisation will not be out of place. As we mentioned in chapter 2, attempts to optimise across complex systems invariably impose constraints on sub-systems. Although this interaction is inevitable, for example one could not construct a practical chemical plant if every unit process were optimised in isolation, one must always be aware of the fact that optimisation implies limitation of degrees of freedom and no change in one part of the system is ever completely undemanding on the rest. A simple and classical case is capital saving by eliminating intermediate storage in sequential processes. In situations in which the output from one unit is the input to the next, if no intermediate storage is provided, and admittedly this may not always be possible, then both units coupled together must work or neither can operate. The dangers of direct coupling are not so great in situations in which one reactor follows another but can be acute when little understood physical processes such as crystallisation are tied directly on to a reactor system, especially in the matter of reaching and then maintaining a steady state of production in the reactor and of separation in the crystalliser.

7.5 The development of the use of mathematical models in process design

As chemical engineering developed as a distinct discipline it saw its special contribution in the study of unit operations – distillation, heat exchange, filtration, etc.; though other topics such as strength of materials and corrosion were also essential components of the chemical engineer's course of study. It was therefore to be expected that the first use to which computers would be put in the design of chemical plant would be in studying unit operations, and so it turned out. Up to about 1970, process designers used computers mainly for specific operations in specific plants and though some generalised programmes were produced for common items such as distillation columns, few were widely used. The main reason for this lack of acceptance was that the programmes were difficult to use, especially in that even trivial changes in parameters, such as the introduction of an alternative procedure for evaluating vapour–liquid equilibrium, could not be made without extensive re-writing of the programme. Even so at least one excellent book was produced as early as 1968 (ref. 16) which predicted many later developments, and there is a useful survey in the *Chemical Engineer* (ref. 17).

It is, of course, the use of the digital computer of which we are writing. Analogue computers were also being employed, especially for modelling chemical reactions, for which the principle of representing processes analogically by electrical circuits which can add, subtract, differentiate and so on, is especially apt. In due course programmes that simulated analogue computing were developed for digital computers which, because of their greater power and flexibility, largely took over the analogue activity.

From about 1970 onwards, the use of the computer gained rapid and wide acceptance in engineering departments, due mainly to the development of generalised programmes and to the availability of easy and quick access to the computer. The use of modular programming methods enabled programmes to be written in a highly flexible way and a number of 'packages' were constructed that enabled models of large complex systems to be produced by non-specialist engineers. Such systems, which are typified by ICI's Flowpack, Monsanto's Flowtran, and others that are produced by computer bureaux, are now used routinely by most chemical companies and some are available to universities for undergraduate teaching. Two useful books that deal with these developments are *Material and Energy Balance Computations* (ref. 18) and *The Application of Mathematical Modelling to Process Design* (ref. 15); the latter gives some useful practical examples.

One problem that still remained was that the data required by the programmes, though available in large part in the literature, were often difficult to find. A current development in some large chemical firms is the setting up of a physical property data bank from which data can be called both by individuals and by the computer through instructions written in to the design programmes.

The importance of recognising the dynamic elements in a situation has been stressed. Current developments in process modelling concern themselves not only with the steady state behaviour of processes, but also with unsteady, transient states, such as may be experienced during start-up or mal-operation. The book *Modelling and Simulation in Chemical Engineering* (ref. 13) is a useful introduction to this area and also contains a useful general discussion on modelling. Hitherto, unsteady state modelling has been the province of the control system designer (ref. 19).

Stochastic modelling, that is to say the modelling of systems in which one or more parameters is determined by a random process, has been used for some time in 'queueing' problems such as determining optimum levels of maintenance staff, optimum stock holdings of finished products and of standby spares, etc. The random processes in these cases are, of course, breakdowns, arrival of orders, and breakdowns again, and the queues are of plant requiring attention (vicariously represented perhaps by plant managers) and of customers. A common type of model which generates a random demand for the solution of such problems is known as the Monte Carlo model (ref. 20). Such models are now being applied to the important area of reliability engineering (ref. 21) which the chemical industry is learning from nuclear power, electronics and aerospace industries.

A growing area of research investigates the logical generation of alternative ways of doing things, constructing possibilities and then screening and evaluating them. (Screening and evaluation will be discussed in the next chapter.) Among the subjects to which this technique has been applied are reaction path synthesis and fault tree analysis. It also has potential application to such tasks as plant layout. This growing field is well introduced in the book *Process Synthesis* (ref. 22).

Progress is being made in the development of systems that are capable of storing in integrated form all the information that is relevant to the construction of a chemical plant. Such a system could integrate with ordering and material control and with interfacing equipment such as graphic displays for detailed mechanical design and by reducing paper work and the number of independent steps should also reduce errors and ensure that all the project design team are kept continuously in the picture.

All these developments have been made possible by the increased power of computers largely as a result of miniaturisation of components and printed circuits and through better means of communicating with the computer. It is probable that the trend will be towards modelling and computing on powerful 'mini' computers in engineers' own offices. These will themselves have extensive capability but will also be linked to data bases and large 'mainframes' that will provide special facilities. The continuous presentation of work in progress to the user, instead of periodic print-outs and graphical presentation will become increasingly important.

All this will perhaps recall to the reader two matters that have been principal themes in this book. Firstly the power of quasi-independent systems, here the minicomputers, which are at last demonstrating their cost-effectiveness, in many applications, over big machines in which the complexity multiplies when one tries to build in facilities to cope with almost any conceivable computation and to run several programmes simultaneously, becomes paralysing. And secondly, the insidious threat of 'technism' once more shows its face. The need to subordinate all activity to good design remains paramount.

8

INNOVATION

8.1 The nature of innovation

The middle chapters of this book are about the use of models to gain insight, to organise thought and to predict, principally in the context of learning and of laboratory research and experimentation. Among the main themes in the first two chapters were the relationship between modelling and design and the use of models to ensure that new artifacts fit well into their task environment. To these aspects of modelling we now return to discuss innovation with especial reference to the development and marketing of new products.

Innovation, as we use the term, is not synonymous with discovery or with invention. Discovery is the uncovering of new knowledge. An invention is a discovery that is recognised as possessing utility. Innovation goes further; it is a process for bringing about changes and specifically those changes that constitute a practicable solution to a problem. Innovation, in our particular context, is a matter not merely of making a discovery nor of appreciating its relevance to a practical situation, but of expressing the new knowledge in a practical form and, as is increasingly important and increasingly difficult, of persuading all those concerned to make or accept the changes that are necessary on their part if the innovation is to take place.

In stressing the importance of innovation as a process (indeed as a conflict in which some assist and some resist, and all perhaps with good reason) we contradict a point of view that, until a few years ago, was clearly in the ascendancy and which even now has its adherents, to wit, that innovation springs essentially from an isolable activity, namely scientific research.

The attributing of overwhelming importance to the discovery has been associated with the influential book *The Sources of Invention* by Jewkes, Sawers and Stillerman (ref. 1) but in our opinion, this model of innovation is misleading because it neglects the inescapable fact that many people other than the discoverers and inventors have a part to play in the process that converts ideas into innovation. This fact is of even greater importance in present times. Not only because the field for innovation is constrained, as we shall see later, but because through a deepening sense of social responsibility, interested parties who hitherto might have been neglected are now heard.

Surprisingly, it appears still to be the case that many a scientist takes Emerson's aphorism that you have only to invent a better mousetrap and the world will beat a path to your door, as truth. It is not. Certainly, successful innovation in a technologically based activity often demands a scientific contribution of a high order, but this alone is by no means enough.

And so the first need in the application of models to innovation is a realistic model of the process of innovation itself; that is to say a model that will direct the scientist's attention to all those factors that are likely to decide the success or failure of his venture; a model that will take cognisance of the imperatives of the process of design, as has been clearly expressed by J. Christopher Jones (ref. 2).

> . . .the designer must be able to predict the ultimate effects of the proposed design as well as specifying the actions that are needed to bring these effects about. The objectives of designing become less concerned with the product itself and more concerned with the changes that manufacturers, distributors, users and society as a whole are expected to make in order to adapt to and benefit from the new design.

Such a conceptual model is attempted in three papers under the title *Patterns of Innovation* by Bradbury, McCarthy and Suckling (ref. 3). The following paragraphs illustrate the essential features of their argument. It is a process in which many people take part and in which the germinal idea, the impetus to innovation, may occur in many places as well as in the laboratory. The impetus may be either the recognition that something could or ought to be done better, but without an initially clear idea of how this might be achieved, or that a discovery creates opportunities by removing previously accepted constraints. For example, the invention of the anaesthetic Halothane, which we will discuss later, followed from the recognition that a better combination of properties was needed in general anaesthetics and several components of this combination of properties were precisely defined before the research began. Alternatively, the impetus might be the desire to find uses for new material or new science. The development of polyethylene is an interesting case. Polyethylene was discovered as the result of basic scientific research with reactions under high pressure but the impetus to developing the manufacturing process was given by the war-time need for good high-frequency insulators in radar equipment. As we all know, the uses to which polyethylene has been put are now many and varied and the cue for innovation has surfaced at various points in the chain, for example the use of polyethylene for shrink wrapping was due mainly to the manufacturers of packaging equipment, that of black polyethylene in agriculture by farmers who perceived its advantages for moisture retention and suppression of weeds.

The use of polyethylene film as a base for wallcoverings arose from attempts to produce a material that would be suitable as a paper substitute in high-quality publications. The venture was not successful in this direction but the material was taken up by a manufacturer of wallcoverings who found not only that it offered new possibilities for design and had a warm, pleasant feel but also that it had great advantages for the consumer in being very easy to apply to the wall and to strip off at the end of its life, and a product was marketed as 'Novamura'. This innovation, like most, needed the support of a number of enabling inventions before it reached the market place, among them novel inks and specially adapted equipment for printing and reeling the polyethylene film.

The discovery of how to make ductile tungsten (ref. 4) is a classic example of an enabling invention. Many materials including tungsten were tried as filaments for incandescent lamps before Edison settled on carbon but, for various reasons, none was satisfactory. Subsequently a 'pressed' filament was developed from tungsten powder mixed with an organic binding material and this proved to be a great improvement over the carbon filament. However, it was still brittle and this meant not only that the lamps had to be handled gingerly but that they could not be used in places where there was much vibration, for example in tramcars. Nevertheless nearly half a million tungsten filament lamps were sold in their first year, 1907. The potential advantages of a ductile filament were obvious and, after a number of false starts, W. D. Coolidge found that, contrary to expectation, pressed tungsten filament which was crystalline and brittle was converted to a fibrous ductile form by drawing it hot through a number of dies. It took two years of painstaking effort to devise a manufacturing process but in 1911 drawn wire lamps were marketed and were so successful that a million dollars worth of pressed filament equipment and bulbs were scrapped. This invention – a process for the manufacture of ductile tungsten – was the final and critical enabling step in the production of cheap incandescent lamps of near universal application.

The process of innovation has a high rejection rate. Very few original ideas are ultimately exploited. In Britain, for example, only 2% of patents survive their full life of sixteen years and in the pharmaceutical industry it is estimated that 3,000 compounds must be synthesised for each one that gets through clinical trials.

Not only is the rejection rate high, but many innovations undergo significant changes of object or method during their development. Indeed it is essential that this should happen since the process that we are describing is the adapting of the initial idea into the real-life situation. The innovatory process is lengthy; eleven years between the initiation of the idea and its commercial realisation is a good working average. Moreover, because of the frequent change in objective or method, the process tends to recycle frequently, that is to say that a product tested in the market place is found defective in some respect or other and this leads to further work in the laboratory and eventually to further evaluation. Most innovations require an input from a variety of technologies and from a variety of organisations and it follows that the cooperation of the necessary supporters is an important factor in the success or otherwise of an innovation.

Lack of awareness of the extent to which many people seemingly remote from the source of an invention can and should influence its progress is a frequent cause of failure of innovation, and an especially fruitful source of friction between academics and industrialists. The academic who offers an invention to industry is often unaware of downstream problems and attributes any tardiness in development to the N.I.H. (not invented here) syndrome or worse. The problem is perennial and may be evidenced and summarised in quotations from two famous inventors and innovators, Baekeland and Gillette.

Baekeland, accepting the Perkin medal, said (ref. 5)

> Experience has taught me that many inventions may look
> very good to an inventor, but some way or another do not
> suit the demands of the public. . .I found to my
> astonishment that people who were proficient in

> manipulation of rubber, celluloid, or other plastics, were
> the least disposed to master the new methods which I tried
> to teach them, or to appreciate their advantages. This was
> principally due to the fact that these methods and the
> properties of the new material were so different in their
> very essence from any of the older processes in which these
> people had become skilled

and Gillette, who spent many years developing his razor, wrote, (refs. 6, 7)

> Every invention that is fundamental in character becomes
> the foundation of endless supplemental ideas and
> inventions that are necessary to its financing, to its
> management, to the development of machinery and
> processes of manufacture, and to its advertising and sales . . .

8.2 Constraints on innovation

It has long been accepted that economic growth and rapid technological change go hand in hand, and many people would argue that technological change is the chief causative factor in the growth of individual economies. This argument is very convincing as there are obvious correlations between societies with high rates of technological change and high economic growth rates. Moreover, the effect has been multiplicative. One technological advance has not only demanded or encouraged other such advances, but it has also freed the human resources to make further advance possible.

Thus, if we plot against time almost any variable which determines our physical environment over the past 2,000 years, the result is always the same; the line is horizontal and on or near the base line for the entire period up to about 100 years ago and then, for certain societies only, the line becomes almost vertical in a period of technologically based, self-sustaining growth. The fact that not all societies have reached this type of growth has caused, and is still causing, a widening of the gap between the rich and the poor countries of the world and much overseas aid from the industrialised countries has been aimed towards helping to provide this 'critical mass' of technological achievement that would enable future growth to be spontaneous.

Estimates of availability of resources, including capital, and of demand that can be derived from macromodels of this kind (by macro-model we mean a generalised description of a complex system) are a basis of some more highly structured models such as the Cambridge input–output model of UK economic growth (ref. 8) which generates quantified descriptions of possible economic futures from sets of inputs that can be varied in order to answer 'what will happen if' type questions.

However, models of self-sustaining growth have received a number of setbacks in recent years. They may well still apply to non-industrialised countries but for the so-called highly developed countries such macromodels have received increasing criticism from the following viewpoints: Firstly limitation of resources must eventually curtail the exponential growth of the recent past. Secondly, it has become increasingly clear that in modern societies the linkages between technology and growth are extremely complex and involve many other variables, some of which are sociological in nature and notoriously difficult to measure or even to foresee. Thirdly, this complexity combined with the speed of technological change is providing a cultural shock which, it is argued by some, is so great that technological change will eventually have to be curtailed in order to avoid social disintegration.

The quarter of a century following the second world war was a period of unparalleled invention in which the chemist played a predominating role. There was major innovation in, among other fields, fibres, drugs and plastics and in the massive scale-up of many industrial processes. The situation is now very different and calls for rethinking of the opportunity and purpose of innovation. A recognition of the problems of pollution and of the potential exhaustion of raw materials, together with a change in attitudes to wealth consumption and shifts in economic power between countries and between classes have eliminated the possibility of the free-for-all innovation which has characterised the last quarter of a century. Moreover there is increasing recognition of the high economic cost of making changes and of providing in wide variety technologically advanced products with very similar properties.

The cost of innovation has indeed risen very greatly. Manpower and the services utilised in research and development are themselves a good deal more expensive. To demonstrate environmental and health acceptability is an increasingly long and costly task, especially in the case of drugs and of products for crop protection, and capital costs for manufacturing plant are escalating very rapidly. Often the prices that consumers are willing to pay for new products are not sufficient to cover the cost of major innovations. In terms of new processes, as we have seen in the previous chapter, the new plant has to produce at less than the variable cost of the old plant since the capital for that plant is written off. Moreover the older plant will be operating relatively efficiently, having considerable learning experience.

It is not only in manufacturing processes that progress along the learning curve helps to entrench existing practice. In any situation in which there is reason to fear environmental or health risk, the well tried and proven product must be difficult to dislodge. The new benefits that are needed to outweigh the risk and the cost of change must be very considerable indeed.

However, the fact that costs are high and resources short at a time when there is a wider expectation than ever of a broadly spread improvement in living standards presents an opportunity for innovation that is directed to the efficient use of resources.

A classical critique that stands in stark contrast to the optimistic projections that held sway for most of this century is the macromodel that was worked out by the Club of Rome (ref. 9) that predicts severe constraints on industrial growth.

Those readers who perceive the overlap of technologically based innovation with the process of design will not suppose that, because most of our examples relate to advanced technology, our argument is irrelevant to the simpler, intermediate technology that may be needed in the less developed countries. The process of design and the practice of modelling are intensely concerned with a realistic, relevant appraisal of whatever problem is in hand. They are highly appropriate techniques for achieving that wider recognition of relevant factors that will be essential if our social and environmental objectives are to be achieved. And though the scientist may, more often than in the past, find it necessary to question the appropriateness of technological sophistication and great size of project (though they will still have their place) he need not discard his scientific models. Rather, as an innovator, he should build on them and expand them to encompass the practical situations in which he wishes his work to be effective.

This book is not the place for a detailed working out of the social, economic and political implications of technological advance which are of course essentially a matter of defining objectives and priorities. Our practical response as scientists can, however, be guided by:

(1) A reconigition of the implications of the 'artificial' elements in technological development and in particular the excessive rate of change that it often imposes on social and physical environments.

(2) The studying of any innovation in a context that is wide enough to embrace the total environment that is likely to be affected.

(3) Identifying the responsibilities of the innovator with those of the designer and by diligently seeking out all the factors that are relevant to the task in hand, ensuring that new products or processes are not misfits in their task environment.

(4) By heeding the warnings of the critics of technism and of pure reason.

8.3 Economic and technological forecasting

The need to make some forecast of future technological change – for a period of 10 to 20 years or so – was discussed in the previous chapter in the context of process development. The same exercise is a necessary component in the assessment of the potential of a new product, but now we must think not only about the technology of product manufacture but also about the technology of the product in use.

It is increasingly important that technology be efficient as well as effective, as the previous section has emphasised, and for this reason economic projections must be developed side by side with the technological; economics being the science of the use of scarce resources. Theoretically, for a given process, a cost or a value can be assigned to all the resources consumed and to all the products produced, and a profitability can be computed essentially as the different between the total cost of inputs and total value of outputs. The increasing concern for the wider social effects of innovation extends the list of elements that should be included in this balance and much research is being done in an attempt to assign a cost to, for example, the adverse effects of noise in order to suggest what it would be worth spending to reduce it. This activity, known as *cost-benefit analysis* (refs. 10, 11) though fraught with difficult subjective value judgments, is likely to become increasingly important.

Everyone is aware of the very low success rate of forecasting for periods of only 3–5 years ahead and it would be easy to take the view that the possibilities of a meaningful forecast of up to 25 years were so limited that it would be wiser to allocate the effort that is used by forecasting to some seemingly more productive purpose. The cost of innovation nowadays is, however, very great and resources are relatively scarce. At several critical stages in the long development someone must take a view and decide whether or not to commit further, often substantial resources to the project. That person will rely on an implicit or explicit model of the long-term future of the innovation. He will have to put the innovation in the context of the future economic, social and technological circumstances which will be relevant not only when the innovation is introduced to the market, but also for many years afterwards.

Such judgments are inescapable. An important step in increasing the probability of correct decisions, and of substituting informed judgment for best guess, is to make the necessary models of the future as explicit as possible.

The different types of explicit modelling that can be useful for decision making in this context fall into the following four groups:

> (1) Models of the overall innovation process which can assist in identifying important parameters and in concentrating available resources on the most critical aspects of the task. Such a model was discussed in section 8.1.
>
> (2) Models that forecast the economic future so that market size, margins and the rate of acceptance of the new product may be better understood.
>
> (3) Models that forecast technological future, so that the risk of technological obsolescence of the new product can be more easily assessed. For many products this determines the life of the product.
>
> (4) Models that will help in selecting projects that are sensible in terms of the predicted social, economic, and technological future, but which also satisfy additional criteria and constraints, for example, that the project is sufficiently profitable, has acceptable cash flow, is consistent with the long-term objectives of the firm and so on. This type of modelling is often termed project appraisal.

These last three types of model will be discussed in sections 8.4, 8.5 and 8.6 respectively. Although the models used in innovation are often treated in the literature as quite distinct activities, they interact strongly with one another.

Mention has been made of the need for the chemist to be part of a multidisciplinary approach to innovation in the chemical industry. Through studying the many interactions of these models we see why the chemist, biologist, physicist or the chemical engineer must work within the context of, and be responsive to, future economic, technological and other developments. The interactions are very much inherent in the task to be performed and are essential elements in fruitful design.

8.4 Economic models

In the last twenty years there has been a great increase in interest in economic modelling of all kinds. There are many interrelated reasons for this, most of which originate from a broad move away from 'laissez-faire' and towards 'planned' economies. 'Planned' in this context has acquired a somewhat specialised meaning, more in the nature of the imposition of constraints on the economy rather than planning in the sense of defining objectives and then deciding on the optimum way of reaching them. The centralised planning of a national economy does offer a number of conceptual benefits, notably in relation to the allocation of scarce resources, but it is a matter of some debate whether or not constraints are becoming so many and so severe that there is no longer the requisite variety in the economic system to allow it to adapt to changing circumstances. One class of constraints in such a situation is the need to obtain agreement from many sections of the nation before any reallocation of resources takes place and this alone can become so inhibiting that the conceptual benefits of centralisation are never realised in practice.

Whatever is the true value of this type of planning, there is little doubt that there is more of it and that it is covering ever wider economic units. The emergence of the institutions of the EEC are good examples of the rapid moves towards 'planning' on a supra-national basis. From the point of view of the innovator, these developments have been a mixed blessing. He suffers from increased limitations on innovation such as higher costs, longer development time and various social pressures but on the other hand the future economic options are better described and better analysed than was the case twenty years ago. Working within such national and supra-national economic plans the innovator may need to make broad economic analyses related to his own particular objectives. Two techniques may be especially useful; *trend analysis* and *input–output analysis*.

8.4.1 *Trend analysis*

Trend analysis is one of the simplest and most useful techniques of economic analysis available to the innovator. It is based on a simple model of the type:

The level of economic activity = An underlying trend
+A 4–5-year economic cycle about this trend + Noise

where the noise term is the residual effect that cannot be explained by the effect of an underlying trend on a regular cycle about the trend. Figure 8.1 shows the actual data for the U.K. index of industrial production (all industries, 1970 = 100) and figure 8.2 a cycle fitted to the data for the years 1964–1974 corrected for the trend. The models can be expressed by the equations:

(*a*) A trend of a modified exponential form with the following constants:

$$\text{index in year } T = 130.4 - 47.3(e^{-0.064\,T})$$

where $T = 1$ in 1964

(*b*) A cycle of period 4.75 years superimposed on the trend of the form:

$$\text{cyclic component of the index} = (1.76 + 0.01\,T)\sin(1.323\,T + 0.436)$$

where $T = 1$ in 1964
and the angles are expressed in radians.

The equations are calculated by a computer programme so written that it will identify which of several types of trend curve best fits the data and then perform the rest of the analysis on this selected curve.

It has turned out that the profitability of a number of UK businesses correlates remarkably closely with the Index of Industrial Production. This connection can be extremely valuable, not only in predicting demand and thus improving production scheduling, but also in computing that crucial factor, cash flow. (Any scientist who is inclined to dismiss cash flow as unimportant should try working for a firm that is in chronic negative cash flow. A satisfactory cash flow is a major

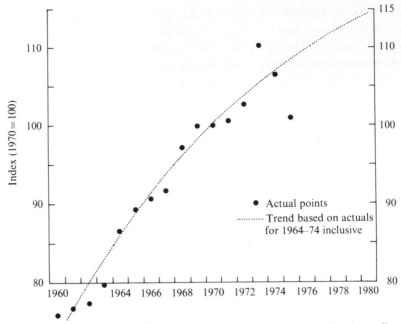

Figure 8.1. United Kingdom index·of industrial production, all industries, seasonally adjusted (data from *Economic Trends*, 1975 Supplement p. 65–7).

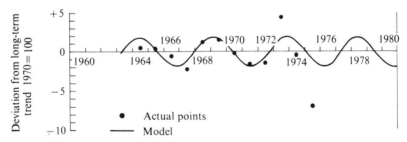

Figure 8.2. United Kingdom index of industrial production all industries (data from *Economic Trends*, 1975 Supplement).

determinant in business stability and growth and therefore of employee prospects.)

The long term trend of demand for some inorganic chemicals such as sulphuric acid and the alkalis sodium hydroxide and carbonate have remained astonishingly stable over many decades, with the effect of two world wars showing as temporary, if substantial, aberrations. Where such a pattern can be relied on, the long term matching of production to demand is relatively simple. Nevertheless, the effect of the overlying cycle, the boom and slump is very significant. Managers tend to be more influenced by short term movements in their economic environment than by long term trends. This is not surprising since the average period in a post is probably one or two cycles. Their own performance and expectation is critically affected by the short term: the long term is uncertain and will in due course assert, if not look after, itself. Nevertheless the planning of investment has until recently taken insufficient account of trade cycles, with the result that investment has been concentrated within the period of rising demand and has fallen away sharply during slump and the chemical construction industry has been alternatively overloaded and under-employed. Attention to both trend and cycle is necessary if both sensible long term strategy and effective, opportunistic tactics are to be employed.

Of course the key question to be answered before relying on a trend is – will it continue? Some trends, such as the rate of growth of the fibres industry at its peak, and of population today, are manifestly unsustainable for long. For many products the sigmoid track familiar to chemists may be expected to apply: slow induction (initial growth), rapid middle course (the years of growth), and then a tail off (maturity). See for example the trend of consumption of titanium dioxide in the U.S. (figure 8.3).

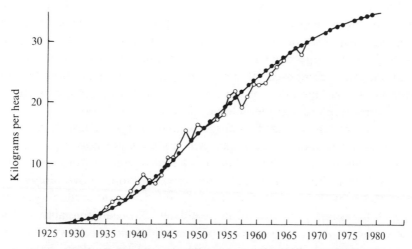

Figure 8.3. U.S. titanium dioxide (1925–1969) consumption per head; ●–● trend, –○–○– observations; curve, Gompertz (from *Ind. Mark. Manage.* **2** (1972)).

Writing as we are early in 1977, the overriding question is, needless to say, how sharp a dislocation in long term trends have the economic upheavals of the last few years brought about?

An excellent paper on the use of trend curves for forecasting the demand for chemicals is that by Harrison and Pearce (ref. 12), which not only describes the underlying theory but gives some actual trend curve analyses based on very long runs of data.

For scientists who have not studied economics but who would like an introduction to the discipline that is both authoritative and apposite to their needs, Shackle's *Economics for Pleasure* (ref. 13) can be strongly recommended, with Loasby's *Choice, Complexity and Ignorance* (ref. 14) as a provocative book that has much to say that is relevant to modelling and to the running of a business.

8.4.2 *Input–output analysis*

Useful though trend analysis is, the relationships which are obtained are essentially statistical in nature. They are not causal models and they are not highly structured. They tend to indicate what is likely to happen if things continued in the future as they have done in the past but they do not enable one to explore what would happen in the future if a specified new set of conditions were to obtain. Input–output analysis provides additional, but causal and well-structured models, for exploring economic environments.

We used three input–output models in the previous chapter; to describe the business in which a manufacturing process must fit, to set out the cash flows of such a business and to represent the balanced material and energy flows into and out of units of the plant. It is clear that they are composed of elements that are related directly and causally to the operation of the process and that they are capable of being structured in considerable detail, if this is desired.

Similar models are used to describe and analyse national, regional or industrial economies in a way which shows not only how money or materials flow through the economy but also what causes these flows to take place. Traditional accounting presentations show neither the flows nor why they occur. For example, the domestic product account (which measures production), consolidates all productive activity for the economy and in the process of consolidation, the flows of intermediate materials between industries are lost. It is often, however, precisely these intermediate flows that are of significance in an economic analysis and the input–output analyst attempts to recreate a complete and balanced network of all intermediate flows between industries, markets or some other suitable subdivision of the economy. A pioneer in the field of input–output modelling was Leontieff (ref. 15) whose work was primarily concerned with the economy of the U.S.A. (A short paper on input–output analysis of business sectors in the U.K. will be found in ref. 16.)

8.5 Forecasting technological change

A large part of technological advance is accounted for by an accumulation of comparatively small changes. An important class of minor technological changes are those improvements to production processes and to products that begin when a new manufacturing plant is first commissioned and continue throughout its working life. These improvements lead to progressively higher output from the same plant, or lower production costs, or both. This learning effect has been studied extensively in a variety of industries and the results have been formulated into generalised models such as the Boston model which enable one to forecast, at an early stage, the expected reduction in cost as experience with the process increases. Perception of the threats and opportunities inherent in potential short-term technological change will depend greatly on the alertness and imagination of the scientists concerned and can be re-inforced by use of models for problem identification of the type that were discussed in chapter 2.

On the other hand, forecasting the occurrence of major technological advances has proved extraordinarily difficult. All forecasting is based on extrapolation of the past into the future. A really substantial change in technology – the break-through – tends to produce abrupt discontinuities in long-established trends and therefore upsets forecasting models, especially those which, as part of the conventional wisdom, are held implicitly in people's minds. Major technological changes which revolutionise an industry are by their nature unexpected. If they were expected then they would be under intensive development and in that situation one is concerned with forecasting not the change but the cost of its successful development and its subsequent rate of diffusion.

8.5.1 *The Delphi technique*

A procedure that has had considerable vogue as an aid to constructing long-term scenarios is the Delphi technique (ref. 17). This is a sort of structured brainstorming in which a number of experts are invited to think out possible future developments, which may be social and political or economic as well as technological and in which their individual views are circulated to the other participants for comment and as a stimulus to further thought. From a series of such iterations, scenarios are built up. The technique makes it possible for a large number of people to participate and facilitates the multi-disciplinary approach that is so conducive of imaginative thinking. 'Workshops', meetings of a group of participants can be used in conjunction with the Delphi exercise and may be particularly useful at the start to define the scope of the study, and at the end to review the results. However, as in all creative thinking it is essential to separate idea generation from idea evaluation; to suspend censorship so that the madness with method in it is not immediately stifled. The Delphi technique can be regarded as a multi-disciplinary comparing of implicit models of the future by people who are in some sense experts in a relvant field, and a projection of possible futures.

8.5.2 *The diffusion of innovations*

In what way and how fast innovations spread is often of social and economic as well as of technological importance and is highly significant for the innovator, be he chemist, engineer, or of whatever discipline. The process is rarely automatic and is often slow. The way in which innovations spread has been extensively studied over the past twenty years and a number of models of the process have been suggested. These models are most often referred to as models of the technological diffusion process. They can be of a very general nature or quite specific. We will refer briefly to a model of each kind.

A good example of the very general model is that due to Rogers and Shoemaker (ref. 18) who have studied how the communication of ideas affects social change. They are concerned with generalisations that are not limited to a particular class of ideas such as say, innovations in the chemical industry, nor to particular social systems, like for example those found in western industrialised countries. Their model of the innovation–decision process is thus very general in nature. It identifies four stages in the process of deciding to accept an innovation, namely, exposure to knowledge that the innovation exists, the development of favourable attitudes towards the innovation, a decision to adopt the innovation, and confirmation by its use. These four stages are developed into a cohesive model and this is then applied to various practical situations. This type of model is closely related to models of the learning process and to the ways in which attitudes are changed.

A very useful feature of Rogers and Shoemaker's book is the bibliographical appendix which summarises generalisations concerning the diffusion of innovations by means of an analysis of several hundred published results. There is also a conventional bibliography which is an excellent starting point in a search for publications dealing with diffusion models for particular products or markets.

Mansfield has devoted most of his research to the second type of model, specific diffusion models. His work on the diffusion of numerical control technology in the American machine tool industry is an interesting example of the development of a highly specific diffusion model (ref. 19). He develops an algebraic model which describes the rate of spread of the innovation and compares the coefficients in the equation with examples from other industries. If one's innovation is a new product that is used in industry then, to the users, it will often be a 'new technology' and in this situation (for example the use of improved fire-fighting chemicals or crop protection products) Mansfield's models can be useful predictors of the rate of growth of demand.

The importance, in achieving technology transfer, of interaction between people as distinct from reliance on organisation, and the circulation of reports is stressed by Bieber (ref. 20).

8.5.3 *The overall economic and technological projection*

In concluding these sections on economic and technological forecasting it may be helpful to note one or two special problems which can arise when the foregoing ideas are applied to particular projects.

Although each project will have its own characteristics which may require special studies, it is nearly always worthwhile to obtain, at a very early stage, an overall appreciation of the long-term economic trends in the community where the innovation will be used. Though very macro in nature, this can often be a very powerful method of checking the feasibility of a proposal at an early stage. For example, suppose that it is known that a given economy has had an average growth rate of 2½% per annum for thirty years. Then any growth rates for a new product which are consistently higher than this would immediately be suspect unless it was clearly shown that the particular market in question would grow consistently faster than the economy as a whole, or that the innovation could penetrate existing markets, or that some kind of substitution would take place. These are simple checks but all too often ignored.

Technological changes must be specific to the area under consideration but an attempt should be made to visualise them in the context of the general economic progress of the community. This then gives the background necessary to carry out a complete project evaluation.

At the end of a project evaluation it is usually necessary to produce figures which show the discounted cash flow rate of return, the forecast return on capital employed, the total new fixed capital required and the cash flow over the life of the project. Many systems also require sensitivity analysis, which helps to show what happens to the cash flow and return if the forecasts on which the proposal is based are subject to specified errors. This type of numerical analysis of projects is covered in a number of text books, some of the best known of which are those by Merrett and Sykes (ref. 21). There is also an excellent inexpensive programmed text available in six volumes which deals with discounted cash flow from first principles and then moves on through most aspects of the numerical evaluation of capital projects (ref. 22).

8.6 Project evaluation

The economic and technological modelling and forecasting that has been briefly introduced in the preceding two sections relates, for the most part, to a broad view of the environment and of the scope for innovation in it. This section and the one that follows focus more closely on the evaluation of the particular project. The activities of *project evaluation* and *product evaluation* overlap in time and are two ways of looking at the same problem. They are presented separately because of the importance of the process of screening products for a particular use and of the relationship of the screen, as a model of the task environment, to the initial target setting and to the ultimate quality control.

Innovation is an avid consumer of resources and is likely to take much longer and cost much more than might be expected. Careful evaluation of the prospects both initially and frequently thereafter is therefore essential. The first step in evaluation may well be to question the need for innovation; to ask whether whatever the organisation is currently doing is appropriate to its task. This question, in common with much of this chapter, is equally relevant for public and private sectors, for the academic as well as for the industrialist, and for the provision of services as well as for the introduction of products. There are many situations in which no major innovation is called for. But if a comparison of current activity with needs as exposed by consideration of task and objectives does throw up ideas for innovation, then a careful evaluation of individual projects will be required. There are many types of model that can be used in project evaluation. In this section we will discuss a type of model, the ranking model, that seeks to establish the factors that are most likely to effect the success of a project and to evaluate quantitatively against them. The reader will find discussion of other models and leads in the books by J. Christopher Jones (ref. 2) and Baines, Bradbury and Suckling (ref. 23).

There are often more innovative leads than can be developed with the resources that are available and choices must be made. Techniques are therefore needed to assist in ranking projects in order of suitability. What contributes to suitability? Clearly this depends both on the objectives and the resources of the innovating organisation. The question of objectives has been briefly touched on in the preceding paragraph and is, of course, a fundamental element in the concept of design that we adopt in this book. In considering resources we must take into account not only what is available for the development phase but also to what extent the new venture will interact, favourably or otherwise, with existing operations; in other words what synergy, positive or negative, can be expected.

Synergy is the working together of two or more factors to produce an effect that is greater than the sum of the effects of each factor working separately. The whole is greater than the sum of the parts. A model in which a firm's potential synergy for specific projects is compared with that of competitors, given in Bradbury *et al.* (ref. 24), considers synergy under people, laboratory and semi-technical equipment, production resources, marketing compatability and customers' requirements and should be consulted for a working out of the concept of synergy as it affects innovation. The importance of synergy in corporate strategy in general is very well presented in Ansoff (ref. 25).

Synergy is rarely to be had for nothing; a price has to be paid in degrees of freedom. Whether it be the utilisation of material that was hitherto waste, the integration of successive stages in a complex manufacturing process or the combining of several distribution systems, to mention only a few from endless possibilities, the interaction that creates the benefit also imposes new constraints on each component. Whether these constraints will give rise to a financial penalty that will cancel out all or part of the expected advantages can only be assessed by careful thinking through of possible constraints over the whole life of the project.

As we have seen, the essential factors to be taken into account in evaluating a project are relevance, cost, benefit and risk. The political, social and personal dimensions are added by asking, for each of the parameters, to whom? To assist in making what are at best difficult choices, a number of techniques have been worked out for ranking projects, and one of these will now be discussed.

A list of the parameters that are thought to be important in a given business situation is drawn up. The strength of each project against each criterion is estimated, sometimes a weighting factor is applied to take account of different degrees of importance that are attached to the parameters. The scores can then be combined to give an overall mark from which projects can be ranked in order of desirability. This method of ranking is dangerously attractive since, though seemingly offering a simple yet quantified selection procedure, it often models too simply and unnecessarily suppresses important detail that, as we shall see, could easily be retained.

Here is a list of parameters that was drawn up for evaluating potential new products:

(1) Can we make it?
(a) is technology available and understood?
(b) what health and environmental hazards are there in manufacture?
(c) is an appropriate work force available?
(d) are raw material and services supply adequate and secure?
(e) are suitable buildings or plant available?

(2) Can we sell it?
(a) are we already in the market?
with what market share?
(b) what is the competitive position; ours, our customers?
(c) whose support/acquiescence will be required for the product's introduction to the market?
(d) what advantages will be offered to the user?

(*e*) what disadvantages are there to the user (need to learn new technology, change equipment, convince his customers)?

(*f*) what health and environmental hazards are there in use?

(3) Can we protect it?

(*a*) by patenting

(*b*) by providing unique back-up service

(*c*) by strong selling and distribution

(4) Profitability

(*a*) fixed capital required

(*b*) working capital required

(*c*) return on capital employed

(*d*) pay-back time (how long after the start up of the plant will the costs have been paid back and profit begin to be earned?)

These parameters constitute one possible model of what may be regarded as the 'task environment' of the project. Used simply as a check-list it can help to ensure that no important factor is neglected and so serve an essential purpose. Some authors have advocated the utilising of such a check-list as the basis of ranking procedure, by subjectively allocating a score for each parameter and adding them to give an overall grading. This procedure is not very helpful and can be greatly improved, firstly by allocating the score according to a set of clearly expressed criteria developed for each parameter, and secondly by not attempting to aggregate all the scores but rather setting them out as a project profile that will highlight strengths and weaknesses. The use of criteria in the ranking process greatly facilitates this. As an example, criteria for requirements 1 *a* and 2 *f* of the above list could be defined as follows, and appropriate sets could be worked out for each factor.

1(*a*) *Overall technical feasibility*

5 All technical problems defined and shown to be soluble.

4 Minor problems of scale-up outstanding.

3 Laboratory process fully established.

2 Minor problems outstanding in laboratory work.

1 Significant problems still to be solved in laboratory work.

2(*f*) *Health and environment factors* (*in use*)
5 Makes positive contribution to reducing
    pollution/hazards.
4 No significant effect on health/environment.
3 Less toxic/hazardous than existing products.
2 Similar level of toxicity/hazard to existing products.
1 Greater hazard in use than existing products.
(o Strong doubt on acceptability to general
    public/industry).

This restructuring has important advantages. The decision whether
to proceed or to abandon a project must, in the last resort, be taken
after a conceptual summation of factors for and against and an act of
judgment. The attaching of a verbal implication to each mark can
inform judgment and provide a basis for further thought, which a simple
number cannot do. The introduction of a zero is important, it expresses
the possibility that an insuperable problem exists. Bradbury and his
co-authors (ref. 24), adopting a similar procedure, assessed each factor
as: no problem, minor problem, major problem, minor threat, major
threat or ignorance. Since an important purpose of research is to reduce
uncertainty, such a classification is a useful guide to where effort
should be directed. These procedures provide an informative profile
that is a far better basis for decision making than a composite and
structureless ranking.

8.7 Product evaluation

In the previous section we discussed project evaluation, with
the object of answering the question – if we had a product that would
meet such and such a need, could we manufacture, distribute and sell
it economically? An important element in this study is, of course, the
definition of the product; what pattern of needs will it be designed to
meet? In this section we will consider first the problem of specifying
the requirements and then of testing compounds to see whether they
meet them, that is to say screening, and finally, that of setting operating
specifications that will ensure that the product that is sold does the
job that it is expected to do. The opportunity for innovation may be
recognised anywhere and sometimes, when the means are ready at
hand, it will be possible to try out the idea quite simply and cheaply.
But when what is called for is a novel product that will demand the
investment of considerable resources in research or implementation,
then a careful analysis of the objective will be in order.

Chemicals are bought for the *pattern* of effects that they produce. When they are used some things must happen and others not, and this is as true for a simple molecule such as sodium carbonate sold by the millions of tons for glass making as it is for the complex antibiotic. What this pattern should be for a particular end use may be definable in advance if the end use is well understood and then the assembling, as a target, of both the set of desirable properties and that of the undesirable can be undertaken, through the listing of misfits or in some other way. Every product must behave appropriately, not simply at the moment of use but from the time when it leaves the factory gate until it has been finally disposed of. That implies, for example, stability in storage. No-one wants to buy a battery that corrodes its casing during a reasonable lifetime, or paint that settles out in the can – a thick sludge of pigment and filler at the bottom and thin unpigmented varnish floating on top. And, potentially far more serious, a fire extinguisher that does not eject its contents when needed is a hazard not a life saver. Similarly, with many products, the problem of non-polluting disposal after use is of increasing importance.

A crucial stage in the developing of a strategy for product innovation is, therefore, the setting of a realistic target, that is to say a statement of the pattern of effects that is wanted. In the case histories at the end of this book there is one example, the development of the anaesthetic 'Halothane' in which it proved possible to specify the target fairly precisely at the outset and a contrasting example, lightweight building blocks in which the full requirements were not recognised initially.

An important input to defining a pattern of needs comes, as would be expected, from market research. Market research aims to find out, by interviewing customers, not only what is wanted but also how much and at what price. It is a highly skilled business that depends very much on seeing the right people and asking questions that elicit useful answers without leading.

It is characteristically the case in the chemical industry that many compounds or mixtures (formulations) can be thought of that might meet a defined need. Very often, for reasons of safety or cost, one cannot test out the candidates in the real life situation. This is most obvious with drugs but it is true to a greater or less extent in most cases. And so we must devise a test, or set of tests, to which potentially suitable products can be submitted that will predict how they will behave in practice – that is to say we have to make a model of the task environment. Since the purpose of this model is to prevent unsatisfactory products reaching the market it has to reject those that will not meet the requirements and is, therefore, usually known as a screen.

In setting up a screen one meets the problems that arise in any other form of modelling; what factors to represent and how to represent them. In product innovation, a well-defined target will be an important initial step. A screen, like any model, needs to be both realistic and economical and the factors that are included must be sufficient but also necessary (refs. 23, 26). In designing a screen and interpreting its output we must have regard to the degree of abstraction inherent in the way in which each factor in the task environment is represented. Some factors can be tested directly. For example, it is quite possible to test fire-fighting chemicals against large fires under practical conditions; one can test static fires of any kind of material and also fires in flowing petrol and the like, and one can test for re-ignition. Although, for reasons of economy, the initial test for extinguishing capability might be very simple, for example introducing a small amount of the sub-stance under test into the gases flowing into a bunsen burner, it is possible to progress to successively larger fires in a thoroughly realistic simulation.

However, not all tests will be so direct. For example, a minimum shelf life of three years for the fire extinguisher would not be an un-reasonable requirement. But must we wait so long before introducing a promising compound or is it possible to find a way of accelerating the effects of potentially disadvantageous factors, for example by programmes of forced vibration and cycling of temperature? Certainly the first thing to do would be to get into store a number of formulations, even if those were not the ones finally adopted, so that they could serve as a basis for the calibration of accelerated tests.

Screens which are not replicas of task environment situations, and that means most screens, are models and therefore incorporate some degree of abstraction. As in all modelling, the limitations on range of relevance that this abstraction imposes must be carefully thought out. Animals used for testing drugs intended for human therapy may be relatively close models, though no animal has proved to be an adequate model for the entire human system. A further degree of abstraction is introduced when isolated organs or tissues are used as screens and still more when biological activity is predicted from physical tests. In all cases a reliable correspondence between the test result and the relevant variety of task environment must be firmly established.

If a product successfully works its way through the screens and negotiates all other hazards on the path to market acceptance then, to control the quality of the manufactured product, a specification will have to be written; that is to say a set of tests that are both necessary and sufficient to ensure that each batch of the product performs in use as it should. Sometimes this will incorporate elements that correspond directly to demands of the task environment, colour and viscosity in paint, adhesion in adhesives, thermal stability in some plastics and so on. Such tests will be needed most frequently when the product is a mixture, the behaviour of which cannot be readily predicted from its chemical constitution. Sometimes, especially with products that are essentially one compound, chemical composition will suffice. In this respect the advances in physical methods of analysis that have taken place over the last twenty-five years have both lightened the task, in providing quick and simple analytical techniques and increased it – by making possible the setting of ever more stringent standards of purity. Indeed the extreme environmentalists demand for 'no detectable *x*' would shackle the economy to the latest advances in analytical sophistication, irrespective of the level of impurity that constitutes a quantifiable and acceptable risk.

In the three stages: desired pattern of effects, set of screens, manufactured product specification, one can see a sequence of related models that if thoroughly thought out and well constructed will do much to ensure that innovation is successful and beneficial in the broadest sense.

8.8 Some case histories

We end this chapter with three short case histories of innovations with which one of the authors has been associated. They exemplify many of the points that have been made in previous sections while, we hope, offering some relatively light reading for the last lap. Halothane and 'Paraquat' as examples of innovation are analysed in some detail in (ref. 3).

8.8.1 *Halothane*

The search for a new inhalant anaesthetic that eventually led to the introduction of Halothane (refs. 27, 28) began with the development of a target through discussions with anaesthiologists and surgeons. Through these discussions it was ascertained that the desirable set of properties of a new anaesthetic, a specification that was not met by any product then available, were as follows.

Firstly and obviously, since what was required was an inhalant, the product must be volatile. Next, anaesthesia must be produced by a relatively low volume percentage of the anaesthetic, say less than 10% by volume, so that a high concentration of oxygen could, if necessary, be administered for supportive purposes. In ideal anaesthesia the patient is unconscious, insensitive to pain and relaxed, and it was desirable that he should reach this stage quickly and smoothly and recover in the same way, without unpleasant after effects such as vomiting or nausea which were common after anaesthesia by some of the then used techniques. And obviously the anaesthetic must not produce any serious side effect, on heart or liver for example. The anaesthetic should be non-inflammable.

A criterion that could easily be missed and an example of the need to thoroughly understand the task environment so as to have the best chance of identifying all potentially serious mismatches, was the requirement not to decompose over soda-lime. This need comes about as follows. In order to conserve expensive anaesthetics it is customary to have the patient breathe in a closed circuit, in which water vapour and carbon dioxide are removed from the expired breath by passage over soda-lime, the gases being then topped up with anaesthetic and oxygen as necessary. In absorbing carbon dioxide and water the soda-lime becomes hot. When, years before, trichloroethylene was first used as an anaesthetic, serious damage was caused to the patient because trichloroethylene was dehydrohalogenated by soda-lime to a toxic chloroacetylene.

Cost would obviously be an important element ultimately, but it was decided to leave this out of consideration initially, since only when the advantages of any compounds that passed the stringent tests that would be necessary were known would it be possible to put a value on them.

There were many compounds that could be thought of as potential anaesthetics. It was obviously impossible to test them in humans without very considerable prior experimentation in animals. However, biological testing is expensive and can use up large amounts of material. There was therefore a big incentive to screen out those compounds that would not meet the chemical and physical requirements before passing anything to the pharmacologist.

Of the desirable properties, one, volatility, is straightforward and physical and can obviously be directly measured. However, the relationship between structure and volatility is fairly well understood and one can therefore use a theoretical model to predict the volatility of new compounds and so avoid wasting time in synthesising materials which are too high-boiling. Likewise, non-inflammability and stability over soda-lime are properties that the chemist can expect to predict fairly readily. In setting up the screen, attention must be given to the conditions as they will be in the operating theatre, and so the anaesthetic must not be inflammable or explosive in high concentrations of oxygen. Likewise the tests over soda-lime had to be conducted in the presence of carbon dioxide and moisture and, to provide a margin of safety, at a higher temperature than was likely to be reached in practice.

This left the biological properties, potency, absence of side effects and adequate rate of induction and recovery to be provided for. At the time when work on Halothane was being done it was not possible to predict with any confidence the relationship between structure and side effects such as, for example, cardiac irregularity, changes in blood pressure, damage to the liver and the like. So there was little theoretical basis for chemical choice in this respect. The simple assumption was made that chemically unreactive compounds that would be excreted unchanged were the least likely to have toxic effects and compounds were synthesised with this in mind. Of course, those compounds that were submitted for test and passed the initial pharmacological screening were evaluated in a battery of tests on a variety of animals which provided information on a very wide spectrum of pharmacological properties.

On the matter of speed of induction and of recovery, which was a question of the dynamics of the transfer of anaesthetic from the breathed air through the lungs into the blood and thence into the brain and the reverse process, there was theory based on the solubility of anaesthetics in the blood that might have helped, but the thorough evaluation of the theoretical possibilities would have been very time consuming and it was considered that this would not be a useful screen.

The screen for anaesthetic potency, on the face of it, would have to be purely biological. In fact it proved possible to predict fairly accurately what the anaesthetic potency of new compounds would be from their physical properties. The basis for this was provided by work done on the narcosis of grain weevils with the object of controlling infestation of bulk grain cargoes. This work (refs. 29, 30) divided volatile chemical compounds into two classes, one of which produced toxic irreversible symptoms and the other non-toxic reversible narcosis.

It showed that, approximately speaking, equal degrees of narcosis (which in this context means the reversible suppression of cellular activity) were produced when volatile compounds were administered at equal thermodynamic activities, i.e. in approximately equal relative saturations. Examination of extant data showed that this generalisation could be applied with reasonable accuracy to any anaesthetic in man and also in mice, the animal which was the primary screen in the biological tests. In fact, it became possible to predict from structural considerations, and later from solubility parameters measured by gas chromatography using polar stationary phases that served in their solubility characteristics as models of the brain, what the anaesthetic potency in test animals and in man would be. This was important not only in focusing attention on that class of compound that was most likely to meet the required specifications, but also in enabling the pharmacologist to economise material which was often very difficult to obtain, because he could begin his experiments with approximately the right concentration, which had been predicted on the basis of gas chromatographic work. Of course, the implication of this theory is that anaesthesia is produced by a physical and not by a chemical mechanism and is, in some way, a matter of differential solubility. This concept profoundly simplified the task because one did not have to search for a complex chemically based model that would explain why such differing compounds as chloroform, ether, acetylene, cyclopropane, trichlorethylene and even xenon, all produced anaesthesia, the appropriate model was simple and physical.

Thus, the research that led to the discovery of Halothane was guided by a set of related models. First, the definition of the target, then models that assisted in interpreting each element of the target in terms of the physical and chemical properties that would be needed. These models greatly increased the probability that the properties of compounds that were selected for synthesis would approximate to the desired pattern. Especially important was Ferguson's model of narcosis not only in the ways that were outlined in the previous paragraph, but also in prompting him to call for a programme of research into anaesthetics when in due course, as a Research Director, he had the opportunity; a programme that ultimately led to the discovery of Halothane.

The case well exemplifies the relationship of the screens, as model of the task environment, to the desired pattern of effects that defined the target. Also that screens can have a high degree of abstraction and yet be extremely effective, provided that realistic connections between screen and use are established. Indeed, the basis of empirical correlations between activity coefficients in polar stationary phases in the gas chromatograph that proved so useful in the Halothane work has been independently developed much further, and on a theoretical basis, by Davies and his co-workers (ref. 31). They found that the anaesthetic potencies of a series of halogenated hydrocarbons could be fitted to a phase distribution model that approximates the activity coefficient as a function of dominant intermolecular inter-actions, essentially the compounds' van der Waals and hydrogen bond donor properties. The authors suggest that theoretical estimates of hydrogen bond donor and acceptor properties in compounds may have a general role to play in evaluating non-specific biological processes. It might be added that in any situation in which an equilibrium concentration might be reached in specific biological processes, the hydrogen bonding properties could play a role in determining what this, possibly critical, concentration would be.

8.8.2 *Lightweight building blocks*

A limestone quarrying company thought up a way of using the small-diameter stone which it currently had to throw away. It would grind it, mix it with sand, foam it, cast it into blocks, cure the blocks and sell them for building walls. Raw material costs were expected to be very low, but this proved to be the first disappointment. Not all waste limestone proved to be satisfactory – and waste material is only cheap when it really is waste. Once you have to sort it or process it, or once someone else finds a use for it, it suddenly becomes very much more costly. However, the dangers were not recognised and the project went ahead.

Not much consideration had been given to the totality of the re-quirements in the task environment, it was thought that if a weather-proof block which was relatively light for its load-bearing capability could be manufactured, then it would be bound to find a niche in building. However, as soon as the block was shown to architects it became clear that sound insulation was an important consideration since the principal use for the block was seen to be in dividing walls. The process was reworked in order to reduce sound transmission and this necessitated the formulation of a denser block and some further loss of the originally perceived cost advantages.

A trial wall was built, the surface was rather rough and needed a layer of plaster. Unfortunately it proved impossible to plaster the wall satisfactorily by normal techniques because the wall was too porous and the plaster dried out from inside far too quickly. Back to the laboratory and the semi-technical plant and this problem too was solved.

In an effort to avoid further mismatches, a thorough study of the expected requirements of building blocks was now undertaken, including such vital economic factors as the costs of transport from the factory to the building site. It was noted with satisfaction that over 50% of the population of England lived within sixty miles of the site in Derbyshire where the factory was to be situated. Unfortunately it was not remarked that practically nobody lived within thirty miles of that spot – the wrong model again.

A technical solution was found to all the problems, including that of ensuring dimensional reproducibility which was vitally important in the accurate and simple construction of walls. A comparative trial was done between the new building blocks and the traditional methods of construction at a building site where the blocks were incorporated into a number of houses. The result was technically and economically satisfactory. The rate at which the walls were built was carefully measured and overall costs of the various types of fabrication were calculated. This showed up the new block to advantage and armed with this information a sales representative set out to visit a number of large builders.

They were not impressed. 'Have you ever been on a real building site? Do you realise how difficult it would be to line up all the materials on time and in the exact quantities and in the right places in the way that you did in your carefully controlled trials? And surely you know that we work largely with casual labour, they have a pretty good idea as to how long it takes to build a wall. They are not going to be motivated as your bricklayers were, taking part in an experiment which interested them, and they will see to it that they get a share in the savings which you predict through your new method and which you have allocated between yourselves and us'. And so the project finally foundered on the social and economic facts of the traditional building site. But a number of very valuable lessons were learnt, lessons which we are now offering to our readers at a substantially reduced cost.

A brief mention of a problem in innovating in golf is relevant and may be amusing. A North American company researched and designed an 'improved' golf ball. It kept within the strict regulations as to form and make, but flew further and was less readily cut by the bad shot than balls already on the market at comparable price. But it did not sell well. It was difficult to find out why but, in the end it became clear that the well-hit shot made the wrong noise, not a clean click, but a fuzzy thud. When you are watching someone driving from the tee you can't pace out the shot to see whether it has that bit extra, so you rely very much on the sound. The defect could not be remedied and the project was abandoned.

8.8.3 *Paraquat*

The history of the discovery and development of the weed killer 'Paraquat' (N,N'-dimethyl-4,4'-bipyridilium) (figure 8.4) and of the related 'Diquat' (N,N'-ethylene-2,2'-bipyridilium) (figure 8.5) illustrates several of the models that have been discussed in this chapter and underlines the need for constant alertness in that, in the case of 'Paraquat', several of the models that were initially used proved to be defective in some important particular (ref. 32). Nevertheless they served a purpose as stepping stones to more realistic and therefore more relevant models.

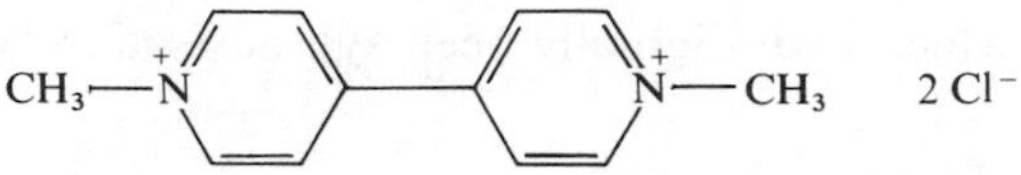

Figure 8.4

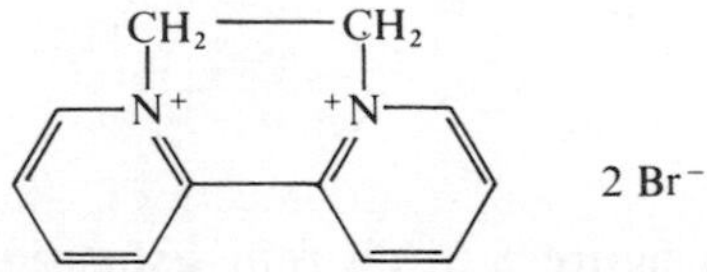

Figure 8.5

The first organic weed killer or herbicide, DNOC (2-methyl-4,6-dinitrophenol) was introduced in 1932, but herbicides did not find significant use until the introduction in 1946 of the 'hormone' weed killers of which MCPA (2-methyl-4-chlorophenoxy acetic acid) is a well known example. These weed killers were selective in that they were principally active against broad-leaf weeds, that is to say against dicotyledons, and they could be applied either to the foliage or to the soil. In 1958 a market survey indicated that the potential for hormone weed killers far exceeded that of any other type of herbicide and ICI's Plant

Protection Group set up screens to test candidate compounds for this type of activity.

The beginning of the bipyridyl story, however, can be traced back to 1947 when a laboratory worker who was operating a screen for wetting agents, that is to say substances that could be added to a mixture of a fungicide or insecticide to ensure that it wetted the leaves and so came into contact with the pest that was to be controlled, found that one of the compounds that was sent to test scorched foliage. Of course, in terms of the objective in hand, phytotoxicity was a complete bar to exploitation but perceptively the adverse effect was reported as potentially useful in a weed killer. The wetting agent was dodecyl-trimethyl ammonium bromide, and it was thought that it scoured the wax from the plant leaves and that this caused scorch. From this hypothesis it was concluded that certain types of quaternary ammonium salts that were good detergents should have the leaf scorching activity. Various quaternary ammonium salts, many of them synthesised for purposes quite unconnected with weed killers, were collected from several parts of ICI and one of them, a substituted bipyridyl, turned out to be highly active both in laboratory screens and in the field.

This compound, which had originally been synthesised in 1950, was

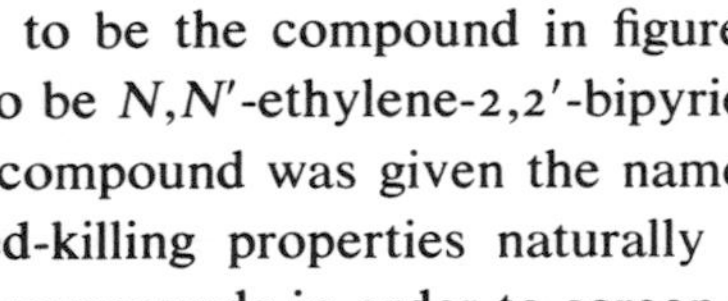

Figure 8.6

thought to be the compound in figure 8.6, on reinvestigation it was found to be N,N'-ethylene-2,2'-bipyridilium bromide (figure 8.5).

This compound was given the name 'Diquat' and the discovery of its weed-killing properties naturally led to the syntheses of other related compounds in order to screen them for phytotoxicity but also to provide additional data on the relationship between structure and activity so that a more realistic model of the mode of action of bipyridyls as phytotoxic agents could be constructed. A number of other bipyridyl salts, among them 'Paraquat', were synthesised and tested. The results led to the conclusion that the damage was caused by the formation from the bipyridylium salts of free radicals which ruptured cell membranes.

Now it was recognised that in this type of compound, stable free radicals were likely to be formed by co-planar systems because in these the delocalisation of an odd electron over the whole molecule by resonance would be greatly facilitated. And this model received support from the fact that 2,2'- and 4,4'-bipyridyls, both of which were co-planar, were more active than 2,4'-bipyridyls which, because of steric hindrance, were not co-planar. Moreover, in 'Diquat', in which the nitrogen atoms were linked by two carbon atoms and which was therefore co-planar, lost activity when the co-planarity was destroyed by breaking the aforesaid carbon–carbon link or by lengthening the carbon chain. This led to a theoretical model on the basis of which phytotoxic potency could be deduced from chemical structure, provided that estimates could be made of the ease with which free radicals would be formed and of their stability. Theoretical models for this purpose are now available but were not at the time of which we are writing. A useful correlation was, however, discovered between the phytotoxicity of a compound and the ease with which it could be reduced, as measured by redox potential. Salts with redox potentials more negative than -500 mV either showed no phytotoxicity or were very slow acting. There were reasons for assuming that the biological system that was affected within the plant possessed a reduction potential of about -380 mV, and on the basis of this value it was possible to calculate the free radical concentration in the plant for an applied volumetric dose that just failed to kill, and this provided a theoretically based comparison. Whereas the sub-lethal volumetric dose differed between bipyridyls by a factor of several hundred, the corresponding free radical concentration differed by a factor of only three. This elimination of apparent differences closely parallels what was found in anaesthesia, in which anaesthetic doses for different compounds expressed as volumetric percentages differ very widely, whereas on the basis of relative saturation dosage is almost a constant. Naturally, when allowance is made for significant parameters the quantitative divergence between the effects produced by various compounds becomes less. In anaesthesia the crucial step was to eliminate differences that were due to concentration effects by using, when expressing dosage, thermodynamic potentials, that is to say relative saturations instead of volumetric concentrations. The insight into the mechanism by which bipyridyls are phytotoxic gave important guidance to the experimentalist because it became apparent that 2,2'- and 4,4'-bipyridyl possessed redox potentials that were very close to the optimum. The way was then clear to devote resources to developing these compounds without continually looking over the shoulder to see whether an outsider from the same stable was coming up on the rails. It is clear that such conclusions are far more likely to be accessible from a theoretical model, which by its nature has a broad field of relevance, than from an empirical model which, as we have suggested in chapter 2, is likely to be, if extrapolated, at best ineffective and at worst dangerous.

'Paraquat' and 'Diquat' were discovered at a time when the accepted model of the market need postulated that the most desirable properties in a weed killer were selectivity and persistence. They turned out to be highly unselective in that they killed green foliage of all kinds, though they would not penetrate woody material. Moreover the bipyridyls were absorbed very rapidly on the soil and therefore almost instantly deactivated. Fortunately, faced with these new possibilities, the developers perceived an alternative model of market need and it was rapidly recognised that bipyridyls possessing a quick knock-down followed by rapid deactivation in contact with soil and, being particularly effective against grasses, might well replace ploughing – the mechanical operation that is undertaken principally to destroy weeds. The desirability of eliminating ploughing by chemical control of weeds had been recognised earlier in the United States where, as is well known, there are large areas which are unsuitable for ploughing either because of rough terrain or because of the danger of soil erosion leading to the dust bowl.

The modelling activity which characterised all stages of the work on 'Paraquat' was also utilised in the development of processes for its manufacture along lines that have been mentioned in the previous chapter and indeed the 'Paraquat' plant at the Pilkington–Sullivan Works of ICI in Widnes was, it is believed, the first complicated chemical plant to be designed from the outset to be operated by on-line computer. To achieve this required the constructing of a detailed model of the chemical processes involved and of the plant's (chemical this time) unit operations. The use of the computer proved to be highly successful from the point of view both of management and of the plant operators.

The bipyridyl that has found greatest commercial use has been the 4,4'-bipyridylium chloride, 'Paraquat'. Although this material is very safe when properly used in the prescribed dilution, it unfortunately became apparent that drinking the concentrate could cause death.

There were a number of accidental deaths, mostly caused through the transfer of 'Paraquat' from the containers in which it was supplied into lemonade bottles and the like from which it was unwittingly drunk by children. (Though irrelevant to the present theme it is perhaps worth passing on the observation made by one toxicologist with whom we discussed 'Paraquat', that he had instilled into his children that they should never drink anything directly from a bottle, because the very act of pouring it into some other receptacle usually provides adequate opportunity to recognise that the stuff is not what you thought it was.) A great deal of work was done on preventing accidental ingestion and studies were undertaken into providing an antidote. Obviously a model of the mechanism by which 'Paraquat' produces its toxic effect would be useful in helping to devise remedial procedures. In most cases of death from 'Paraquat' there is extensive pulmonary damage characterised by oedema, haemorrhage and in later stages fibrosis. Except when extremely large amounts were taken, signs of pulmonary damage were not usually seen for several days after ingestion and death might not occur for several weeks. 'Paraquat' was shown to have similar effects in the lungs of rats which were therefore used as a test organism. An energy-dependent accumulation in the rat lung was demonstrated *in vitro*, and this observation accounted for the fact that after oral administration of 'Paraquat' to rats, the lung accumulated 'Paraquat' to a concentration greatly in excess of that in the blood plasma. It was found that the slow onset of lung damage was associated with the absorption of a high concentration of 'Paraquat' in the gut and its subsequent slow release into the plasma, from which the lung accumulated it. When measures were taken to remove free 'Paraquat' from the gut, then the concentration of 'Paraquat' in the plasma was considerably reduced. This led to the suggestion (ref. 33) that early treatment designed to remove 'Paraquat' from the gut might prove to be an effective antidote. Other possibilities were also pursued vigorously.

It is perhaps not inappropriate that our attempt to contribute to the more effective practice of chemistry, through the use of models and effective design, should end with a case history of a product that combines outstanding benefits in use with real dangers in misuse.

REFERENCES

Chapter 1

1 Mislow, Kurt, *Introduction to Stereochemistry*, Benjamin (1966)
2 Hemmerich, Peter, *Model Studies on Flavin-Dependent Oxidoreduction, Vitamins and Hormones* (1970) pp. 467–73
3 Hesse, Mary B., *Models and Analogies in Science*, University of Notre Dame Press (1970)
4 Kuhn, Thomas S., *The Structure of Scientific Revolutions*, 2nd edn, University of Chicago Press (1970)
5 Mihram, G. Arthur, The modelling process, Transactions on systems, *Men and Cybernetics*, Vol. SMC–2 No. 5 (1972) pp. 621–9
6 Bruner, Jerome S., *Towards a Theory of Instruction*, Harvard U.P. (1967): (*a*) p. 10, (*b*) p. 28, (*c*) p. 25
7 Ackoff, Russell L., *Scientific Method: Optimizing Applied Research Decisions*, Wiley (1962) p. 109
8 Beer, Stafford, *Decision and Control, The Meaning of Operational Research and Management Cybernetics*, Wiley (1966): (*a*) p. 100
9 Beer, Stafford, *Platform for Change*, Wiley (1975)
10 Baines, A., Bradbury, F. R. and Suckling, C. W., *Research in the Chemical Industry*, Elsevier (1969)
11 Rivett, Patrick, *Principles of Model Building, The Construction of Models for Decision Analysis*, Wiley (1972)
12 Jones, J. Christopher, *Design Methods, Seeds of Human Futures*, Wiley (1970): (*a*) p. 42, (*b*) p. 22
13 Franks, Roger G. E., *Modelling and Simulation in Chemical Engineering*, Wiley (1972)
14 Pounds, W. F., The process of problem finding, *Industrial Management Review*, **11** No. 1 (1969)
15 Loasby, Brian J., An analysis of decision processes, *R & D Management*, **4** No. 3 (1974): (*a*) pp. 149–55, (*b*) p. 153
16 Simon, Herbert A., *The Sciences of the Artificial*, MIT Press (1969) p. 6
17 Davies, Duncan S., Banfield, Tom and Sheahan, Ray, *The Humane Technologist*, Oxford University Press (1976)
18 Miller, George A., The magical number seven plus or minus two: Some limits on our capacity for processing information, *Psychological Review*, **63** (1956) pp. 81–97

19 Ashby, W. Ross., *An Introduction to Cybernetics*, Chapman and Hall (1956)

Chapter 2

1 Bruner, Jerome S., *Towards a Theory of Instruction*, Harvard U.P. (1967)
2 Morris, William T., *Management Science, A Bayesian Introduction*, Prentice-Hall (1968)
3 Jenkins, G. M. and Youle, P. V., *Systems Engineering*, C. A. Watts & Co. Ltd (1971)
4 Ackoff, Russell L., Towards a system of systems concepts, *Management Science* **17** No. 11 (1971) 661–71
5 Raybould, E. B., Personal communication
6 Kelly, George A., *A Theory of Personality, The Psychology of Personal Constructs*, Norton (1963) p. 6
7 Denbigh, K. G., Hicks, M. and Page, F. M., *Trans. Farad. Soc.* **44** (1948) p. 479
8 Prigogine, I., *Etude Thermodynamique des Phénomènes Irréversibles*, Paris (1947)
9 Banks, Barbara E. C., Thermodynamics and biology, *Chem. Brit.* **5** (1969) 614
Pauling, Linus, Structure of high-energy molecules, *Chem. Brit.* **6** (1970) 468
Huxley, A. F., Energetics of muscle, *Chem. Brit.* **6** (1970) 472
Wilkie, Douglas, Thermodynamics and biology, *Chem. Brit.* **6** (1970) 477
Ross, R. A. and Vernon, C. A., A reply to Douglas Wilkie, *Chem. Brit.* **6** (1970) 539
10 Sypher, Wylie, *Literature and Technology, The Alien Vision*, (1968): (*a*) p. 7, (*b*) p. 27, (*c*) p. 25, (*d*) p. 9
11 Simon, Herbert A., *The Sciences of the Artificial*, MIT Press (1969) p. 16
12 Alexander, Christopher, *Notes on the Synthesis of Form*, Harvard University Press (1971): (*a*) p. 85
13 Jones, J. Christopher, *Design Methods, seeds of human futures*, Wiley (1970) p. 19
14 Davies, O. L. and Goldsmith, P. L. (Eds.) *Statistical Methods in Research and Production*, 4th Edn, Hafner (1972)
15 Beer, Stafford, *Decision and Control, The Meaning of Operational Research and Management Cybernetics*, Wiley (1966): (*a*) p. 97, (*b*) p. 166
16 Mihram, G. Arthur, *Operational Research Quarterly*, **23** No. 1 (1972) 17–29
17 Fishman, G. S., and Kiviat, P. J., *Digital Computer Simulation: Statistical Considerations*, Rand Corporation (RM–5387) (1967)
18 Box, George E. P. and Drapper, Norman R., *Evolutionary Operation: A Method for Increasing Industrial Productivity*, Wiley (1969)

19 Loasby, Brian J., *Choice, Complexity and Ignorance, An enquiry into economic theory and the practice of decision making*, Cambridge University Press (1976): (*a*) p. 43, (*b*) p. 50

20 Michel, Gérard, *Jacques Ibert* Paris (1967) p. 95

21 Eliot, T. S., *Murder in the Cathedral*, Faber and Faber (1935)

22 Churchman, Charles W., *Challenge to Reason*, McGraw Hill (1968)

23 Katz, D. and Kahn R. L., *The Social Psychology of Organisations*, Wiley (1966) p. 27

24 Checkland, P. B., Towards a systems-based methodology for real-world problem solving, *J. Sys. Eng.* **3** No. 2 (1972) 87–115

25 Checkland, P. B., A systems map of the universe, *J. Sys. Eng.* **2** No. 2 (1971)

26 Vickers, G., *Value Systems and Social Process*, Basic (1968), and Vickers, G., *Freedom in a Rocking Boat*, Allen Lane (1970)

Chapter 3

1 Roberts, J. D. and Caserio, M. C., *Modern Organic Chemistry*, Benjamin (1967) p. 257

2 Gould, E. S., *Mechanism and Structure in Organic Chemistry*, Holt Reinhart and Winston Inc. (1959) p. 199

3 Williams, J. E., Stang, P. J. and Schleyer, P. von R., *Annual Rev. Phys. Chem.* **19** (1968) 531

4 March, J., *Advanced Organic Chemistry: Reactions, Mechanisms and Structure*, McGraw Hill (1968) p. 231

5 Katritzky, A. R. and Topsom, R. D., *J. Chem. Educ.* **48** (1971) 427

6 Breslow, R., *Organic Reaction Mechanisms*, 2nd Edn, Benjamin (1969) p. 1

7 Brown, H. C., *Accounts Chem. Res.* **6** (1973) 377

8 Bethell, D. and Gold, V., *Carbonium Ions – An Introduction*, Academic Press (1967) pp. 232, 271

9 Raber, D. J., Bingham, R. C., Harris, J. M., Fry, J. L. and Schleyer, P. von R., *J. Amer. Chem. Soc.* **92** (1970) 5977

10 Hammett, L. P., *Chem. Rev.* **17** (1935) 125

11 Taft, R. W., *J. Amer. Chem. Soc.* **79** (1957) 1045 Taft, R. W. and Grob, C. A., *J. Amer. Chem. Soc.* **96** (1974) 1236

12 Westheimer, F. H., in *Steric Effects in Organic Chemistry*, Newman, M. S. (Ed.), Wiley (1970) p. 524

13 Shorter, J., *Quart. Rev.* **24** (1970) 433

14 Gleicher, G. J. and Schleyer, P. von R., *J. Amer. Chem. Soc.* **89** (1967) 591

15 Pearson, R. G., *Chem. Brit.* **3** (1967) 103

16 Klopman, G., and Hudson, R. F., *Tetrahedron Letters* (1967) 1103

17 Schrödinger, E., *Ann. Phys. Leipzig* **79** (1926) 361, 489
18 Murrell, J. N., Kettle, S. F. A. and Tedder, J. M., *Valence Theory*, Wiley (1965): (*a*) pp. 15, 132, (*b*) p. 79, (*c*) pp. 38–45, (*d*) p. 27, (*e*) p. 340, (*f*) p. 203
19 Streitwieser, A., *Molecular Orbital Theory for Organic Chemists*, Wiley (1961): (*a*) p. 7, (*b*) p. 10, (*c*) p. 20, (*d*) p. 22, and references
20 Jørgensen, W. C. and Salem, L., *The Organic Chemist's Book of Orbitals*, Academic Press (1973)
21 Heitler, W. and London, F., *Z. Physik* **44** (1927) 455
22 Richards, W. G. and Horseley, J. A., *Ab Initio Calculations for Chemists*, Clarendon (1970)
23 Smith, M., *Rodd's Chemistry of Carbon Compounds*, 2nd Edn, **IIA** (1967) 29
24 Pople, J. A., *Accounts Chem. Res.* **3** (1970) 217
25 Moore, W. A., *Physical Chemistry*, 5th Edn, Longmans (1972): (*a*) p. 449, (*b*) p. 457, (*c*) p. 461
26 Fuoss, R. and Krauss, C., *J. Amer. Chem. Soc.* **55** (1933) 21
27 Pauson, P. L., *Quart. Rev.* **9** (1955) 391
28 Jørgensen, C. K., *Structure and Bonding* **1** (1966) 23
29 Thompson, D. W., *Structure and Bonding* **9** (1971) 27
30 Latimer, W., *Oxidation Potentials*, 2nd Edn, Prentice-Hall
31 Cotton, F. A. and Wilkinson, G., *Advanced Inorganic Chemistry*, 3rd Edn, Wiley (1972) p. 898

Chapter 4

1 Moore, W. J., *Physical Chemistry*, 5th Edn, Longmans (1972): (*a*) p. 364, (*b*) p. 363, (*c*) p. 381, (*d*) p. 378, (*e*) p. 464
2 Eyring, H. and Polanyi, M., *Z. Physik. Chem.* **B12** (1931) 279
3 Horsley, J. A., Jean, Y., Moser, C., Salem, L., Stevens, R. M. and Wright, J. S., *J. Amer. Chem. Soc.* **94** (1972) 279
4 Jencks, W. P., *Catalysis in Chemistry and Enzymology*, McGraw Hill (1969): (*a*) pp. 243–9, (*b*) p. 170
5 Rouvray, D. L., *American Scientist* **61** (1973) 729
6 Corey, E. J., *Quart. Rev.* **25** (1971) 455
7 Russell, C. A., *Chemical Bonding*, Leicester University (1971) p. 126
8 Dewar, M. J. S. and Lo, D. H., *J. Amer. Chem. Soc.* **93** (1971) 7201
9 Lipscomb, W. N., *Pure and Applied Chemistry* **29** (1972) 495
10 Olah, G. A., *Chem. Brit.* **8** (1972) 281
11 Dougherty, R. C., *J. Amer. Chem. Soc.* **93** (1971) 7187
12 Robinson, R., *J. Chem. Soc.* (1917) 876
13 Eliel, E. J., *Stereochemistry of Carbon Compounds*, McGraw Hill (1962) pp. 16ff., p. 124
14 Bordwell, F. G., *Accounts Chem. Res.* **3** (1970) 281
15 Zimmermann, H. E., *Accounts Chem. Res.* **5** (1972) 393

16 Arnett, E. M., Jones III, F. M., Taagepera, M., Beauchamp, J. L., Henderson, W. G., Holz, D. and Taft, R. W., *J. Amer. Chem. Soc.* **94** (1972) 4724

17 Ahrland, S., *Structure and Bonding*, **5** (1968) 118

18 Lincoln, S. F., *Coord. Chem. Rev.* **6** (1971) 309

19 Wertz, D. L. and Kruh, R. K., *Inorg. Chem.* **9** (1970) 595

20 Hughes, E. D., Juliusburger, F., Masterman, S., Topley, B. and Weiss, J., *J. Chem. Soc.* (1935) 1525

21 Allinger, N. L., Tai, C. J. and Wu, F. T., *J. Amer. Chem. Soc.* **92** (1970) 579

22 Hughes, E. D. and Ingold, C. K., *J. Chem. Soc.* (1935) 244

23 Grunwald, E. and Winstein, S., *J. Amer. Chem. Soc.* **70** (1948) 846
Grunwald, E., Winstein, S. and Jones, H. W., *J. Amer. Chem. Soc.* **73** (1951) 2700

24 Burgess, D. J. and Price, M. G., *J. Chem. Soc. A* (1971) 3108

25 Taube, H. and Gould, E. S., *Accounts Chem. Res.* **2** (1969) 321

26 Cordes, E. H. and Dunlap, R. B., *Accounts Chem. Res.* **2** (1969) 329

27 Bender, M. L., *Mechanisms of Homogeneous Catalysis from Protons to Proteins*, Wiley (1971)

28 Hucknall, D. J., *Selective Oxidation of Hydrocarbons*, Academic Press (1974) p. 126

29 Hammond, G. S., *J. Amer. Chem. Soc.* **77** (1955) 334

30 Cram, D. D. and Abdelhafez, F. A., *J. Amer. Chem. Soc.* **74** (1952) 5828

31 Morrison, J. D., *Survey of Progress in Chemistry* (1966) 147
Mosher, H. S. and Morrison, J. D., *Asymmetric Organic Reactions*, Prentice-Hall (1971)

32 Pople, J. A., *Accounts Chem. Res.* **3** (1970) 217

33 Dewar, M. J. S., *Fortschr. der Chemische Forschung*, **23** (1971) 1

34 Dewar, M. J. S., *Chem. Brit.* **11** (1975) 97

35 Dewar, M. J. S. and Haselbach, E., *J. Amer. Chem. Soc.* **92** (1970) 590

36 Dewar, M. J. S., Nahlovska, Z. and Nahlovsky, D., *Chem. Commun.* (1971) 1377

37 Dewar, M. J. S., *J. Amer. Chem. Soc.* **97** (1975) 6591

38 Pople, J. A., *J. Amer. Chem. Soc.* **97** (1975) 5306

39 Hehre, W. J. *J. Amer. Chem. Soc.* **97** (1975) 5308

40 Williams, J. E., Stang, P. J. and Schleyer, P. von R., *Annual Rev. Phys. Chem.* **19** (1968) 531

41 Allinger, N. L., Tribble, M. T., Miller, M. A. and Wertz, D. H., *J. Amer. Chem. Soc.* **93** (1971) 1637

42 Bingham, R. C. and Schleyer, P. von R., *J. Amer. Chem. Soc.* **93** (1971) 3189

43 Woodward, R. B. and Hoffmann, R., *The Conservation of Orbital Symmetry*, Verlag Chemie (1970)
44 Gill, G. B. and Willis, M. R., *Pericyclic Reactions*, Chapman and Hall (1974)
45 Evans, M. G., *Trans. Farad. Soc.* **35** (1939) 824
46 Dewar, M. J. S., *The Molecular Orbital Theory of Organic Chemistry*, McGraw Hill (1969)
47 Pearson, R. G., *Accounts Chem. Res.* **4** (1971) 152
48 Fukui, K., *Accounts Chem. Res.* **4** (1971) 57
49 Dewar, M. J. S., Kirschner, S. and Kollmar, H. W., *J. Amer. Chem. Soc.* **96** (1974) 5240
50 Sullivan, J. H., *J. Chem. Phys.* **46** (1967) 73
51 Bloch, K. in *Biological and Chemical Aspects of Oxygenases*, Bloch, K. and Hayaishi, O. (Eds.), Maruzen (1966) p. 23

Chapter 5
1 Woodward, R. B., *Pure and Applied Chemistry* **17** (1968) 529
2 Ireland, R. E., *Organic Synthesis*, Prentice-Hall, (1969): (*a*) chapter 1, (*b*) p. 81
3 Corey, E. J., *Pure and Applied Chemistry* **14** (1967) 9
4 Sargent, G. D., *Tetrahedron Letters* (1970) 4359
5 Sondheimer, F., *Accounts Chem. Res.* **5** (1972) 81
6 Griffiths, J. and Sondheimer, F., *J. Amer. Chem. Soc.* **89** (1969) 7518
7 Sondheimer, F. and Beeby, P. J., *J. Amer. Chem. Soc.* **94** (1972) 2128
8 Battersby, A. R., *Pure and Applied Chemistry* **14** (1967) 117
9 Baines, A., Bradbury, F. R. and Suckling, C. W., *Research in the Chemical Industry*, Elsevier (1969): (*a*) p. 104
10 Hansch, C., *Accounts Chem. Res.* **2** (1969) 232
11 Parke, D. V., *Chem. Brit.* **8** (1972) 102
12 Hendrickson, J. B., *J. Amer. Chem. Soc.* **93** (1971) 6847, **97** (1975) 5763, 5784
13 Marks, G. S., *Heme and Chlorophyll*, Van Nostrand Reinhold (1969)
14 Altman, L. J., Kowerski, R. C. and Rilling, H. C., *J. Amer. Chem. Soc.* **93** (1971) 1782
15 Corey, E. J. and Achiwa, K., *Tetrahedron Letters* (1969) 1837, 3257
16 Cava, M. and Mitchell, M. J., *Cyclobutadiene and Related Compounds*, Academic Press (1967)
17 Emerson, G., Watts, L. and Pettit, R., *J. Amer. Chem. Soc.* **87** (1965) 131, Reeves, P., Hénery, J. and Pettit, R., *J. Amer. Chem. Soc.* **91** (1969) 5888, Reeves, P., Devon, T. and Pettit, R., *J. Amer. Chem. Soc.* **91** (1969) 5890
18 Lin, C. Y. and Krantz, A., *Chem. Commun.* (1972) 1111

19 Corey, E. J., *Quart. Rev.* **25** (1971) 455

20 *Organic Reactions*, **1–23** Wiley 1942–

21 Corey, E. J., Weinshenker, N. M., Schaaf, T. K. and Huber, W., *J. Amer. Chem. Soc.* **91** (1969) 5675

22 Stork, G. and Tomasz, M., *J. Amer. Chem. Soc.* **84** (1962) 310

23 Whaley, W. M. and Gonvindachari, T. R., *Organic Reactions*, **6** (1951) 74

24 Wittig, G., Weigmann, H-D. and Schlosser, M., *Chem. Ber.* **94** (1961) 676

25 Corey, E. J. and Yamamoto, H., *J. Amer. Chem. Soc.* **92** (1970) 226

26 Reucroft, J. and Sammes, P. G., *Quart. Rev.* **25** (1971) 137

27 Stork, G., *Pure and Applied Chemistry* **17** (1968) 383

28 Johnson, W. S., Vredenburg, W. A. and Pike, J. E., *J. Amer. Chem. Soc.* **82** (1960) 3409

29 Wiesner, K., Poon, L., Jirkowsky, I. and Fishman, M., *Canad. J. Chem.* **47** (1969) 433

30 Corey, E. J., Mitra, R. B. and Uda, H., *J. Amer. Chem. Soc.* **86** (1964) 485

31 Woodward, R. B., *Pure and Applied Chemistry* **25** (1971) 283

32 Woodward, R. B., *Pure and Applied Chemistry* **33** (1973) 145

33 Mislow, K., *Introduction to Stereochemistry*, Benjamin (1966): (*a*) p. 42, (*b*) p. 9

34 Walton, A., *Progress in Stereochemistry* **4** (1969) 335

35 Büchi, G., Kulsa, P., Ogasawara, K. and Rosati, R. L., *J. Amer. Chem. Soc.* **92** (1970) 999

36 Eliel, E. L., *Stereochemistry of Carbon Compounds*, McGraw Hill (1962) p. 236

37 Buckingham, D. A., Mason, S. F., Sargerson, A. M. and Turnbull, K. R., *Inorg. Chem.* **5** (1966) 1649
Blount, J. F., Freeman, H. C., Sargerson, A. M. and Turnbull, K. R., *Chem. Commun.* (1967) 324

38 Buckingham, D. A. and Marzilli, L. G., *Inorg. Chem.* **6** (1967) 1042

39 Berkoff, C. E., *Quart. Rev.* **23** (1969) 372
Dahm, K. H., Röller, H. and Trost, B. M., *Life Sci.* **7** (1968) 129

40 Brand, J. C. D. and Scott, A. I., *Technique of Organic Chemistry* **11** (1963) 110

41 Lindskog, S., *Structure and Bonding* **8** (1970) 153

42 Bentley, T. W. and Johnstone, R. A. W., *Advances in Physical Org. Chem.* **8** (1970): (*a*) p. 151, (*b*) p. 157

43 Budziciewicz, H. C., Djerassi, C. and Williams, D. H., *Mass Spectroscopy of Organic Compounds*, Holden Day (1967)

44 Abraham, R. J., *The Analysis of High Resolution NMR Spectra*, Elsevier (1971)

45 Jackman, L. M. and Sternhell, S., *Applications of NMR in Organic Chemistry*, Pergamon (1969) p. 281

46 Sharpless, K. B., and van Tamelen, E. E., *J. Amer. Chem. Soc.*
91 (1969) 1848

47 Elmsley, J. W., Feeney, J. and Sutcliffe, L. H., *High Resolution
NMR Spectroscopy*, Vol. 2, Pergamon (1966) p. 969

48 Stork, G. and Shultz, A. G., *J. Amer. Chem. Soc.* **93** (1971)
4074

49 Eglinton, G. in *Physical Methods in Organic Chemistry*,
Schwartz, J. C. P. (Ed.), Oliver and Boyd (1964) p. 35

Chapter 6

1 Wilson, R. S. and Mertes, M. P., *Biochemistry* **12** (1973) 2879
Pagolotti, A. L. and Santi, D. V., *Biochemistry* **13** (1974) 456

2 Sigman, D. S. and Mooser, G., *Annual Reviews Biochem.* **44**
(1975) 895

3 Robinson, R., *J. Chem. Soc.* (1917) 876, *The Structural
Relationships of Natural Products* Clarendon (1955)

4 Bu'lock, J. D., *The Biosynthesis of Natural Products*, McGraw
Hill (1965): (*a*) p. 361
Hendrickson, J. B., *The Molecules of Nature*, Benjamin (1965)

5 Battersby, A. R., *Quart. Rev.* **15** (1959) 259
Battersby, A. R. and Harper, B. J., *Proc. Chem. Soc.* (1959) 152

6 Ogston, A. G., *Nature* **162** (1948) 963
Dixon, M. and Webb, E. C., *Enzymes*, Longmans (1964) p. 6

7 Gray, C. J., *Enzyme Catalysed Reactions*, Van Nostrand Reinhold
(1971) p. 173

8 Perutz, M. F., *Europ. J. Biochem.* **8** (1969) 455

9 Kassera, H. B. and Laidler, K. J., *Canad. J. Chem.* (1969) 4031,
Blow, D. M., *Accounts Chem. Res.* **9** (1976) 145

10 Baines, A., Bradbury, F. R. and Suckling, C. W., *Research in the
Chemical Industry*, Elsevier (1969) p. 44

11 Robinson, R., *J. Chem. Soc.* (1917) 762

12 van Tamelen, E. E., *Fortschr. Chem. Org. Naturstoffe* **19** (1961)
242

13 Taylor, W. I. and Battersby, A. R., (Eds.) *Oxidative Coupling of
Phenols*, Marcel Dekker (1970)

14 Schwartz, M. A. and Holton, R. A., *J. Amer. Chem. Soc.* **92**
(1970) 1090

15 Mulheirn, L. J. and Ramm, P. J., *Chem. Soc. Reviews* **1** (1972)
259

16 Stork, G. and Burgstrahler, A. W., *J. Amer. Chem. Soc.* **77**
(1955) 5068
Eschenmoser, A., Ruzicka, L., Jeger, O. and Arigoni, D., *Helv.
Chim. Acta* **38** (1955) 1890

17 Johnson, W. S., *Accounts Chem. Res.* **1** (1968) 1

18 Sharpless, K. B., *J. Amer. Chem. Soc.* **92** (1970) 6999

19 Johnson, W. S., Gravestock, M. B. and McCarry, B. E., *J. Amer.
Chem. Soc.* **93** (1971) 4332

20 Cornforth, J. W., Cornforth, R. H., Donninger, C. and Popjak, G., *Proc. Roy. Soc.* **B163** (1966) 492

21 Baldwin, J. E., Hackler, R. E. and Kelly, D. P., *J. Amer. Chem. Soc.* **90** (1968) 4758
Blackburn, G. M., Ollis, W. D., Plackett, J. D., Smith, C. and Sutherland, I. O., *Chem. Comm.* (1968) 186

22 Altman, L. J., Kowerski, R. C. and Rilling, H. C., *J. Amer. Chem. Soc.* **93** (1971) 1782, van Tamelen, E. E. and Schwartz, M. A., *J. Amer. Chem. Soc.* **93** (1971) 1780
Rilling, H. C., Poulter, C. Dale, Epstein, W. W. and Larson, Brent, *J. Amer. Chem. Soc.* **93** (1971) 1783

23 Bruice, T. C. and Benkovic, S. J., *Bioorganic Mechanisms,* Vol. 1, Benjamin (1966): (*a*) p. 212, (*b*) p. 225
Hartley, B. S., *Annual Rev. Biochem.* **29** (1960) 45

24 Bender, M. L. and Zerner, B., *J. Amer. Chem. Soc.* **83** (1961) 2391

25 Doherty, D. G. and Vaslow, F., *J. Amer. Chem. Soc.* **74** (1952) 931
Bender, M. L. and Turnquest, B. W., *J. Amer. Chem. Soc.* **79** (1957) 1652

26 Bruice, T. C. and Schmir, G. L., *J. Amer. Chem. Soc.* **79** (1957) 1663

27 Bender, M. L., *Chem. Rev.* **60** (1960) 53

28 Jencks, W. P., *Catalysis in Chemistry and Enzymology,* McGraw Hill (1969): (*a*) p. 67, (*b*) p. 1, (*c*) p. 393

29 Bruice, T. C. and Sturtevant, J. M., *J. Amer. Chem. Soc.* **81** (1959) 2860

30 Sheehan, J. C., Bennett, G. B. and Schneider, J. A., *J. Amer. Chem. Soc.* **88** (1966) 3455

31 Klapper, M. H., *Progress in Bioorganic Chemistry* **2** (1973) 55

32 Wagner, T. E., Chen-Jung Hsu and Pratt, C. S., *J. Amer. Chem. Soc.* **89** (1969) 6366

33 Cordes, E. H. and Dunlap, R. B., *Accounts Chem. Res.* **2** (1969) 329

34 Van Etten, R. L., Sebastian, J. F., Clowes, G. A. and Bender, M. L., *J. Amer. Chem. Soc.* **89** (1967) 3242, 3253
Cramer, F. and Hettler, H., *Naturwissenschaften* **54** (1967) 625

35 Bender, M. L., Vander Jagt, D. L. and Killian, F. L., *J. Amer. Chem. Soc.* **92** (1970) 1016

36 Blow, D. M., Birktoft, J. J. and Hartley, B. S., *Nature* **221** (1969) 337
Bentley, R., *Molecular Asymmetry in Biology,* Vol. 2, Academic Press (1969) p. 248

37 Lake, A. W., Lowe, G. N., Photaki, I. and Baradakos, V., *J. Chem. Soc.* (*C*) (1968) 1860

38 Kendrew, J. C., Dickerson, P. E., Strandberg, B. E., Hart, R. G., Davids, D. R., Phillips, D. C. and Shore, V. C., *Nature* **185** (1960) 422

39 Antonini, A., Caputo, A. and Rossi Fanelli, A., *Advan. Protein Chem.* **19** (1964) 73

40 Gibson, Q. H. in *Biological Oxidations*, Singer, T. P. (Ed.), Wiley (1968) p. 379

41 Falk, J. E., *Porphyrins and Metalloporphyrins*, Elsevier (1964)

42 Boyer, E. and Schretzmann, P., *Structure and Bonding* **2** (1967) 181, Olivé, G. Henrici and Olivé, S., *Angew. Chem. Int. Edn* **13** (1974) 29

43 McClellan, W. R. and Benson, R. E., *J. Amer. Chem. Soc.* **88** (1966) 5165

44 Wang, J. H., *J. Amer. Chem. Soc.* **80** (1958) 3168

45 Wang, J. H., *Science*, **167** (1970) 25

46 Castro, C. E. and Davies, H. F., *J. Amer. Chem. Soc.* **91** (1969) 5405

47 Dickerson, R. E., Kapka, M. L., Weinzierl, J., Varnum, J., Eisenberg, D. and Margoliash, E., *J. Biol. Chem.* **242** (1967) 3015

48 Malström, B. G., *Pure and Applied Chemistry*, **24** (1970) 393

49 Birnbaum, E. R., Gomez, J. E. and Darnall, D. W., *J. Amer. Chem. Soc.* **92** (1970) 5287

50 Griffith, O. H. and Waggoner, A. S., *Accounts Chem. Res.* **2** (1969) 17

51 Kayre, F. J. and Reuben, J., *J. Biol. Chem.* **246** (1971) 6227

52 Vallee, B. L., Riordan, J. F. and Coleman, J. E., *Proc. Nat. Acad. Sci. U.S.* **49** (1963) 109

53 Williams, R. J. P. and Nennard, A. E., *Transition Metal Chem.* **2** (1966) 122

54 Lindskog, S., *Structure and Bonding* **8** (1970) 153

55 Vallee, B. L. and Williams, R. J. P., *Proc. Nat. Acad. Sci. U.S.* **59** (1968) 498

56 Palmer, G. R., Bray, R. C. and Beinert, H., *J. Biol. Chem.* **239** (1964) 2657, 2673

57 Huang, T. J. and Haight, G. P., *J. Amer. Chem. Soc.* **92** (1970) 2336

58 Gunsalus, I. C., Pederson, T. C. and Sligar, S. G., *Annual Reviews Biochem.* **44** (1975) 377

59 Parke, D. V., *Chem. Brit.* **8** (1972) 102

60 Fieser, L. F. and Fieser, M., *Reagents for Organic Synthesis*, Wiley **1** (1967) 472

61 Lemmon, R. M., *Chem. Rev.* **70** (1970) 95

62 Ring, D., Wolman, Y., Friedmann, N. and Miller, S. L., *Proc. Nat. Acad. Sci. U.S.* **69** (1972) 765

63 Ponnamperuna, C., Lemmon, R. M., Mariner, R. and Calvin, M., *Proc. Nat. Acad. Sci. U.S.* **49** (1963) 737

64 Hodgson, G. W. and Baker, B. L., *Nature* **216** (1967) 29

65 Steinman, G., Kenyon, D. H. and Calvin, M., *Biochim. Biophys. Acta* **124** (1966) 339

Sulston, J., Lohrman, M., Orgel, L. E., Schneider-Bernlöhr, H., Miles, H. Todd and Weinman, B. J., *J. Mol. Biol.* **40** (1969) 227

66 Mildvan, A. S., in *The Enzymes*, Boyer (Ed.) 3rd Edn, Academic Press 2 (1970) 445

67 Abeles, R. H., Hutton, R. F. and Westheimer, F. H., *J. Amer. Chem. Soc.* 79 (1957) 712

68 Cornforth, J. W., Popjak, G., Donninger, C. and Schröpfer Jr, G., *Biochem. Biophys. Res. Commun.* 9 (1962) 371
Arigoni, D. and Eliel, E. L. *Topics in Stereochemistry* 4 (1969) 127
Sund, H., in *Biological Oxidations*, Singer, T. P. (Ed.), Interscience (1968) p. 603

69 Kosower, E. M., *The Enzymes*, 2nd Edn, Academic Press 3 (1960) 171

70 Shifrin, S., *Biochim. Biophys. Acta* 81 (1964) 205

71 Meister, A., *Biochemistry of Amino-Acids*, 2nd Edn, Academic Press (1965) p. 376

72 Snell, E. E., Braunstein, A. E., Severin, E. S. and Tirchinsky, Xu M., *Pyridoxal Catalysis – Enzymes and Model Systems*, Wiley (1968)

73 Lenhert, P. G. and Hodgkin, D. C., *Nature* 192 (1961) 937

74 Smith, E. L., *Vitamin B_{12}*, 3rd Edn, Methuen 1966

75 Schrauzer, G. N., *Accounts Chem. Res.* 1 (1968) 97

76 Schrauzer, G. N., Deutsch, E. and Windgassen, R. J., *J. Amer. Chem. Soc.* 90 (1968) 2441

77 Schrauzer, G. N. and Sibert, J. W., *J. Amer. Chem. Soc.* 92 (1970) 1022

78 Abeles, R. H., Frey, P. A., Essenberg, M. K. and Kerwar, S. S., *J. Amer. Chem. Soc.* 92 (1970) 4488

79 Bruice, T. C. and Benkovic, S. J., *Bioorganic Mechanisms*, Benjamin, 2 (1966) 181, 354, 386

80 Katchalski, E., Sela, M., Silman, H. I. and Berger, A., *The Proteins*, Academic Press 2 (1964): (a) p. 405, (b) p. 519, (c) p. 546

81 McDonald, C. C. and Phillips, W. D., *J. Amer. Chem. Soc.* 91 (1969) 1513

82 Milstein, S. and Cohen, L. A., *Proc. Nat. Acad. Sci. U.S.* 67 (1970) 1143
Hillery, P. S. and Cohen, L. A., *Chem. Commun.* (1972) 403

83 Neet, K. E., Nanci, A. and Koshland Jr, D. E., *J. Biol. Chem.* 243 (1968) 6392

84 Bruice, T. C., Brown, A. and Harris, D. G., *Proc. Nat. Acad. Sci. U.S.* 68 (1971) 658

85 Wang, J. H., *J. Amer. Chem. Soc.* 77 (1955) 4715

86 Stadtman, E. R., The Enzymes, 3rd Edn, Academic Press 1 (1970) 397

87 Koshland, D. E., The Enzymes, 3rd Edn, Academic Press 1 (1970) 341

88 Atkins, D. E., The Enzymes, 3rd Edn, Academic Press 1 (1970) 461

89 Monod, J., Changeux, J.-P. and Jacob, F., *J. Mol. Biol.* **6** (1963) 306

Monod, J., Wyman, J. and Changeux, J.-P., *J. Mol. Biol.* **12** (1965) 88

90 Datta, P., *Science* **165** (1969) 556
91 Perutz, M. F., *Nature* **228** (1970) 726
92 Breslow, R., *Chem. Soc. Rev.* **1** (1971) 553
93 Suckling, C. J. and Suckling, K. E., *Chem. Soc. Rev.* **3** (1974) 387

Chapter 7

1 Checkland, P. B., Towards a systems-based methodology for real-world problem solving. *J. Sys. Eng.* **3** No. 2 (1972) 87–115
2 Shackle, G. L. S., *Economics for Pleasure*, 2nd Edn, Cambridge University Press (1968)
3 Redwood, Heinz, *Mind your own Business (How to succeed in business with charts)*, Leviathan House (1974)
4 Lockyer, K. G., *An introduction to Critical Path Analysis*, 3rd Edn, (1970)
5 Boston Consulting Group, *Perspectives on Experience* (1972)
6 Baines, A., Bradbury, F. R. and Suckling, C. W., *Research in the Chemical Industry*, Elsevier (1969)
7 Ball, D. F. and Pearson, A. W., The future size of process plant, *Long Range Planning* **9** No. 4 (1976) pp. 18–27
8 Lawley, H. G., Size up plant hazards this way, *Hydrocarbon processing* **55** No. 4 (1976) 247 on

Gibson, S. B., The design of new chemical plant using hazard analysis, *I. Chem. E. Symposium Series*, No. 47
9 de Vries, M., *Solving River Problems by Hydraulic and Mathematical Models*, Delft Hydraulics Laboratory, Polish Academy of Sciences. Institute of Hydro-Engineering, Gdansk (1969)
10 Schoemaker, H. J., *Some Pitfalls in Scaling Hydraulics Models*, Publication No. 79 of the Delft Hydraulics Laboratory, Polish Academy of Sciences Institute of Hydro-Engineering, Gdansk (1969)
11 Thomas, P. J., Rate data in process development, *Brit. Chem. Eng.* **15** No. 4 (1970) 214–15
12 Andrew, S. M., Digital computers in the design of complete chemical processes, *Brit. Chem. Eng.* **14** No. 8 (1969) 1057–1062

Andrew, S. M., Computer modelling and optimisation in the design of a complete chemical process, *Trans. Inst. Chem. Engrs.* **47** (1969) T79–T84
13 Franks, Roger G. E., *Modelling and Simulation in Chemical Engineering*, Wiley (1972)
14 Savas, E. S., *Computer Control of Industrial Processes*, McGraw Hill (1965)

15 Rose, L. M., *The Application of Mathematical Modelling to Process Development and Design*, Applied Science Publishers (1974)

16 Rudd, D. F. and Watson, C. C., *Strategy of Process Engineering*, Wiley (1968)

17 Various, Computer applications in chemical engineering, *Chem. Eng.* (1968) 345–53

18 Henley, E. J. and Rosen, E. M., *Material and Energy Balance Computations*, Wiley (1969)

19 Stainthorpe, F. P. and Benson, R. S., Computer aided design of process control systems, *Chem. Eng.* (1974) 531–5

20 Davies, O. L. and Goldsmith, P. L. (Ed.), *Statistical Methods in Research and Production*, 4th edn, Hafner (1972)

21 Green, A. E. and Bourne, A. J., *Reliability Technology*, Wiley (1972)

22 Rudd, D. F., Powers, G. J. and Sirola, J. J., *Process Synthesis*, Prentice-Hall (1973)

Chapter 8

1 Jewkes, J., Sawers, D. and Stillerman, R., *The Sources of Invention*, Macmillan (1958)

2 Jones, Christopher, *Design Methods, Seeds of Human Futures*, Wiley (1970)

3 Bradbury, F. R., McCarthy, M. C. and Suckling, C. W., Patterns of innovation Pts 1–3, *Chemistry and Industry* (1972) pp. 22, 105, 195

4 Howell, J. W. and Schroeder, Henry, *History of the Incandescent Lamp*, The Maqua Company (1927)

5 Baekeland, L. H., Practical life as a complement to university education, *J. Ind. Eng. Chem.* 8 (1916) 18–90

6 Baldwin, George B., The invention of the modern safety razor: A case study of industrial innovation, *Explorations in Entrepreneurial History* 4 (1951–2) 73–102

7 Bezanson, Anne. The invention of the safety razor: A case study of industrial innovation, Explorations in Entrepreneurial History 4 (1951–2) 193–8

8 Department of Applied Economics, Cambridge, England. A Programme for Growth, 6 vols. (1962–1965) London

9 Meadows, Donella H., Meadows, Dennis L., Randers, Jorgen and Betirens, William W., *The Limits to Growth*, Universe (1972)

10 Mishan, E. J., *Cost Benefit Analysis*, Praeger (1971)

11 Kendall, M. G., *Cost Benefit Analysis*, English University Press (1971)

12 Harrison, P. J. and Pearce, S. F., The use of trend curves as an aid to market forecasting, *Ind. Mark. Manage.* 2 (1972)

13 Shackle, G. L. S., *Economics for Pleasure*, 2nd edn, Cambridge University Press (1968)

14 Loasby, Brian J., *Choice, Complexity and Ignorance, An enquiry into economic theory and the practice of decision making*, Cambridge University Press (1968)
15 Leontief, W., *Input–Output Economics*, Oxford University Press (1966)
16 *Economic Trends* No. 270, HMSO London, April 1976
17 Helmer, Olaf. *Analysis of the Future, The Delphi Model*, Boston Consulting Group (1967)
18 Rogers, Everett M. and Schoemaker Ffloyd, F., *Communication of Innovations*, Free Press (1971)
19 Mansfield, Edwin and Rapoport, John, *Research and Innovation in the Modern Corporation*, Norton (1971)
20 Bieber, Hermann., Technology transfer in practice. *IEEE Transactions on Engineering Managements* **EM-16** (1969) 144–7
21 Merrett, A. J. and Sykes, Allan, *The Finance and Analysis of Capital Projects*, 2nd edn, Longmans (1973)
22 I.C.I. Ltd., *Assessing Projects, A Programme for Learning*, 2nd edn, 6 vols. Methuen (1970): (1) Discounted Cash Flow, (2) A Publishing Project, (3) A Manufacturing Project, (4) Sensitivity Analysis, (5) Risk Analysis, (6) Replacing Equipment
23 Baines, A., Bradbury, F. R. and Suckling, C. W., *Research in the Chemical Industry*, Elsevier (1969)
24 Bradbury, F. R., Gallagher, W. M. and Suckling, C. W., Qualitative aspects of the evaluation and control of research and development projects, *R & D Management* **3** No. 2 49–57 (1973)
25 Ansoff, H. I., *Corporate Strategy*, McGraw Hill (1965)
26 Bradbury, F. R., The Theory of Screening. *Proc. 10th Br. Weed Control Conf.* (1970) 986–96
27 Suckling, C. W., Some chemical and physical factors in the development of halothane, *British Journal of Anaesthesia* **29** (1957) 466
28 Suckling, C. W., Raventos, J., Spinks, A. and Johnstone, Michael, The development of halothane. *Manchester University Medical School Gazette* **37** No. 2 (1958) 53–63
29 Ferguson, J., The use of chemical potentials as indices of toxicity. *Proc. Roy. Soc. Series B* **127** No. 848 (1939) 387–404
30 Ferguson, J. and Pirie, H., The toxicity of vapours to the grain weevil, *Annals of Applied Biology* **35** No. 4 (1948) 532–50
31 Davies, R. H., Bagnall, R. D. and Jones, W. G. M. A quantitative interpretation of phase effects in anaesthesia. *Int. J. Quantum Chem.: Quantum Biology Symp.* No. 1 (1974) 201–12
32 Bradbury, F. R., McCarthy, M. C. and Suckling, C. W., Patterns of innovation Pt 3, The bipyridyl herbicides, *Chemistry and Industry* (1972) 195–200
33 Smith, L. L., Wright, A., Wyatt, J. and Rose, M. S. Effective treatment for paraquat poisoning in rats and its relevance to treatment of paraquat poisoning in man, *British Medical Journal* **4** (1974) 569–71

ADDITIONAL REFERENCES

Modelling, design and philosophy

Lossee, J., *A Historical Introduction to Philosophy of Science*, Oxford University Press (1972, reprinted 1977)

Lakatos, I. and Musgrave, A. (Eds.) *Criticism and the Growth of Knowledge*, Cambridge University Press (1970)

Simon, H. A., *Models of Discovery*, D. Reidel (1977)

Archer, L. B., *Design Awareness and Planned Creativity in Industry*, Design Council of Great Britain; Office of Design, Department of Industry, Trade and Commerce, Ottawa (1974)

Bender, E. A., *An Introduction to Mathematical Modelling*, Wiley (1978)

Molecular orbital theory

Dewar, M. J. S. and Dougherty, R. C., *The Perturbation Molecular Orbital Theory of Organic Chemistry*, Plenum (1975)

Fleming, I., *Frontier Orbitals and Organic Chemistry Reactions*, Wiley (1976)

Biological chemistry (chapter 6)

p. 212–17 (Hydrolytic enzymes), Bender, M. L. and Komiyama, M., *Bioorganic Chemistry*, van Tamelen, E. E. (Ed.), Vol. 1, p. 19, Academic Press (1977)

p. 223 (Cytochrome-*c*), Cusanovich, M. A., *ibid.*, Vol. 4, p. 117

p. 235 Staunton, J., *Primary Metabolism*, p. 66, Oxford University Press (1977)

p. 231 (Hydroxylase models), Dewar, C. A., Suckling, C. J. and Higgins, R., *J. Chem. Res.* (1979), (S) 335–7, (M) 3801–30.

Non-classical ions

Brown, H. C., *The Non Classical Ion Problem*, Plenum Press (1977)

INDEX